Semantic Web

Michael Workman
Editor

Semantic Web

Implications for Technologies and Business Practices

Editor
Michael Workman
Advanced Research and Development
Security Policy Institute
Melbourne
Florida
USA

ISBN 978-3-319-16657-5 ISBN 978-3-319-16658-2 (eBook)
DOI 10.1007/978-3-319-16658-2

Library of Congress Control Number: 2015938599

Springer Cham Heidelberg New York Dordrecht London

Printed on acid-free paper

Springer International Publishing AG Switzerland is part of Springer Science+Business Media (www.springer.com)

Preface

Who can conceive of an organization that does not involve information and systems? Information created and used in organizations reflects all the intellectual property, competitive intelligence, business transactions and records, and other strategic, tactical, and operating data for businesses and people. Regardless of industry, people in organizations today need some understanding of how to utilize these technology and information resources. Nevertheless, information (or cognitive) overload has become such a problem as to become a cliché. This seems even more the case if people work in some form of "knowledge work," a term coined by Peter Drucker referring to one who works primarily with information or one who develops and uses knowledge in their work.

According to the Gartner Group and Aberdeen Research, spending on information systems technologies exceeded the $ 2.26 trillion mark per year worldwide in 2012. Yet research has shown that as much as 25–30 % of information technology goes unused after purchase, and of those technologies used, only a fraction of the available features are utilized. Why is so much money wasted on technologies that are later shelved? Research has shown that the primary reasons for this disuse are that people frequently do too much work for the computer rather than the other way around—this is the so-called ease-of-use problem; and that once people are able to access their information, the information is often irrelevant or obsolete—the so-called usefulness problem.

The wasteful spending on technologies is indicative also of other insidious conditions: technologies are not helping people make better decisions, solve problems better, make better plans, or take better courses of action—leading to unbounded costs associated with lost productivity, lost strategic opportunities, tactical missteps, lost revenues, unnecessary expenses, and the myriad of other problems that result from this waste.

In recent years, there has been an explosion of disruptive technologies. Disruptive technologies are those that radically change a computing paradigm. Without the proper understanding of how to design, implement, or even utilize them, these are likely to fall short of their promise. An area of particular interest for our purposes includes the recent developments in semantic systems and Web 3.0 applications that can respond to situations and environments and events. These technologies do not merely serve up passive displays of information for human consumers to digest, but rather they are intelligent systems that are capable of assisting human beings with the creation of meaning and drawing inferences to improve human performance.

We hope you will enjoy this volume on semantic technologies!

Michael Workman, Ph.D.
Chief Research Scientist
Security Policy Institute

Contents

Contributors

Pavani Akundi Department of Computer Science and Engineering, Southern Methodist University, Dallas, TX, USA

Jorge Bernad Department of Computer Science & Systems Engineering, University of Zaragoza, Zaragoza, Spain

Kent D. Bimson Intelligent Software Solutions, Inc., Colorado Springs, CO, USA

Department of Electrical & Computer Engineering, The University of New Mexico, Albuquerque, NM, USA

Carlos Bobed Department of Computer Science & Systems Engineering, University of Zaragoza, Zaragoza, Spain

Fernando Bobillo Department of Computer Science & Systems Engineering, University of Zaragoza, Zaragoza, Spain

Andrew W. Crapo GE Global Research, Niskayuna, NY, USA

Ángel Luis Garrido Department of Computer Science & Systems Engineering, University of Zaragoza, Zaragoza, Spain

Steven Gustafson GE Global Research, Niskayuna, NY, USA

Karim Hadjar Multimedia Department Chairman, Ahlia University, Manama, Bahrain

Eric Hillerbrand Knowledgegrids, Inc., Wilmette, IL, USA

Richard D. Hull Intelligent Software Solutions, Inc., Colorado Springs, CO, USA

Sergio Ilarri Department of Computer Science & Systems Engineering, University of Zaragoza, Zaragoza, Spain

Eduardo Mena Department of Computer Science & Systems Engineering, University of Zaragoza, Zaragoza, Spain

Daniel Nieten Red Lambda, Inc., Longwood, FL, USA

Adrian Paschke AG Corporate Semantic Web, Institute for Computer Science, Freie Universität Berlin, Berlin, Germany

Daniel Riding Eastern Florida State College, Melbourne, FL, USA

Kia Teymourian Department of Computer Science, Rice University, Houston, TX, USA

Raquel Trillo-Lado Department of Computer Science & Systems Engineering, University of Zaragoza, Zaragoza, Spain

Michael Workman Advanced Research and Development, Security Policy Institute, Melbourne, FL, USA

Roberto Yus Department of Computer Science & Systems Engineering, University of Zaragoza, Zaragoza, Spain

Chapter 1
Introduction to This Book

Michael Workman

In this book, we hope to raise some provocative questions as much as we want to answer questions about semantic and Web 3.0 technologies. We will begin by introducing cognitive and sociological foundations for why semantic technologies are superior to their predecessors. We then present some contrasting views about specific techniques, followed by some specific examples. We conclude with a look at the state of the art in semantic systems and their implications for businesses and technologies.

We begin a primer on a few key semantic technologies to orient our concepts and vocabulary—i.e., what we mean by semantic systems. Let us start with the notion that semantic systems include dynamic, self-describing models (and a language for constructing these models), semantic resolution among disparate information sources (called ontologies), and the ability to discover these models (Skyttner 1996). We will also broach the idea that semantic systems also subsume social and biologically inspired systems. With these features in place, the addition of semantic brokering and reasoning/inference capabilities may complete a solution for semantic integration, which is a primary goal of many, if not most, of semantic and Web 3.0 technologies.

1.1 Resource Description Framework

There is a trend in moving away from programmed logic to dynamically generated and interpreted logic within the World Wide Web Consortium (W3C) definitions for semantic technologies (Berners-Lee et al. 2001). For example, we are creating new forms of markup, including the Resource Description Framework (RDF; Lassila and Swick 1999), to enrich information and enable intelligent systems. This evolution came about because there is a need for a more advanced approach to information

M. Workman (✉)
Advanced Research and Development, Security Policy Institute, Melbourne, FL, USA
e-mail: workmanfit@yahoo.com

M. Workman (ed.), *Semantic Web*, DOI 10.1007/978-3-319-16658-2_1

and description logics than was possible with HTML or even XML. While RDF is an XML of sorts, that is to say, RDF is based on XML, it is an attempt to make better use of metadata (data about data) by extending into relationships of the data. For example, when you are going to type up a research paper, you might first search the information related to the topic. You might use a search engine which sifts through metadata looking for keywords or combinations of keywords (without regard to the keyword relationships) that might match. RDF, on the other hand, establishes relationships that go beyond keywords and basic knowledge representations (Miller 1998). RDF imposes structure that provides for the expression of relationships needed for the first step toward semantic systems (which is the ability to associate things with their functions or meanings). RDF consists of resources, properties, and statements. A resource is the metadata that defines the given RDF document and is contained at a specified Uniform Resource Identifier (URI). A property is an attribute of the resource such as author or title. A statement consists of the combination of a resource, a property, and its attribute value. These form the "subject," "predicate," and "object" of an RDF statement, such as in the RDF statement:

```
<rdf:Descr iption about='http://www.my.com /RDF/home.html'>

    <Author>Mike Workman</Author>

    <Home-Page rdf:resource ='http://www.my .com' />

</rdf:Description>
```

In this example, we can see that a document contains a URI, which is very much like the URL we type into our browsers. It redirects the program reading it to that resource, which will likely be another document.

The assertion is that the document at URI: "http://www.my.com/RDF/home.html" is authored by Mike Workman whose homepage is at URI: http://www.my.com. We could tie more documents together via other URI to form networks of associations. Since a property is an attribute of a resource, any person or even a program can create them, and since RDF statements are essentially a form of XML, they can be dynamically produced (generated) and read (consumed). With RDF statements, we make assertions, such as:

1. Mike Workman is a software engineer.
2. Mike Workman teaches network multimedia.
3. Mike Workman has an office at the Library and Information Sciences (LIS) building.

The dynamic and associative aspects of RDF essentially come from four attributes (Bray 2003): (1) Resources can be defined independently, (2) RDF can be canonized for exchange, (3) RDF enables persistent triples (subject–predicate–object), and (4) RDF enables heritable properties. We can see the flexibility this provides, and it is through this flexibility that the RDF enables universal linkages seen in Fig. 1.1.

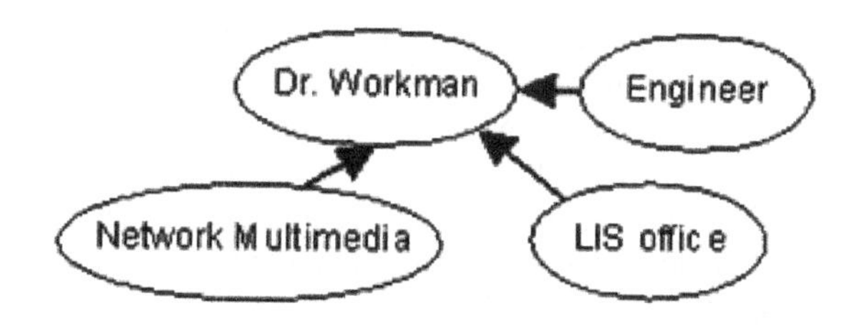

Fig. 1.1 RDF relational associations

1.2 Ontology Markup

It is one thing to infer from relationships, but things fall into classes, and more specifically, objects. From this concept, we may get the sense that RDF can deal fairly well with some aspects of controlled vocabularies such as an antonym–synonym problem, but it does not really help much with the disparate aggregated constituent problem, such as in the case where a "Customer" is not necessarily a "Customer," or a "Teacher" is not a "Teacher" in a processing or even a schematic sense—because their attributes or constituents differ. For example, a set of highly normalized relational database tables that refer to "Teacher" may consist of teacher names, employee numbers, and the atomic pieces that define each teacher, whereas a software object that refers to "Teacher" may consist of not only names but also departments, subject matter taught, rank, and so forth. A reference to one is not necessarily a reference to the other. Equating, differentiating, and resolving these entities go beyond their relationships.

The DARPA Agent Markup Language (DAML) is a markup language that enables computers to draw conclusions based on their constituents because DAML organizes RDF into classes. Thus, in addition to the ability to dynamically specify relationships among entities, descriptive logics such as DAML enable systems to draw conclusions using RDF. If an application is given new information, it can provide additional information based on DAML statements. In other words, DAML statements enable applications to draw conclusions or inferences from other DAML statements.

The Ontology Inference Layer (OIL) is a syntactic encoding language for creating ontologies by allowing humans or software objects (referred to as agents) to markup RDF for knowledge representation and inference. It combines modeling primitives from programming languages with the formal semantics and reasoning services from description logics. Combined DAML+OIL provides the constructs needed to create ontologies (a body of related concepts) and mark up RDF in a machine-readable format, enabling a rich set of object-oriented capabilities (Fresse and Nexis 2002), such as the ability to define not only subclass–superclass relationships but also rules about them such as whether they are disjoint, unions, disjunctions, or have transitivity, along with the imposition of a range of restrictions on when specific relationships are applicable.

Collections of RDF/DAML+OIL can be assembled into even more complex relationships that enable disparate semantic resolution via agents' exchange of ontologies (Fresse and Nexis 2002). This has two dimensions: (1) The unequivocal sharing of semantics so that when the ontology is deployed, it can be interpreted in a consistent manner and (2) when the ontology is viewed by an agent (person or software object) other than the author, it helps to ensure that the intent or meaning of the author is clear. Deriving from these technologies, the Web Ontology Language, or OWL, evolved. The OWL has extended beyond DAML+OIL, and provides a good example of how there is actually a family of web ontology markup languages in the marketplace to choose from.

1.3 Agent Frameworks

Whereas RDF and ontology markup languages and processors have advanced into practice in many systems utilized by businesses, agent frameworks have not yet matured at the same pace. Nevertheless, some of the more recent advances in these technologies have illustrated their potential and viability, particularly in synthetic systems, robotics, and mobile ad hoc networks (Workman et al. 2008). There have been discussions about the merits of agent versus agentless systems in conventional technologies, but in a semantic systems sense, the concept of an agent is much more robust than stationary collector entities that reside on devices, such as in the case of network or application monitors. Semantic agents form a social network and have varying degrees of "problem-solving" capabilities such as setting goals and monitoring progress toward goal completion.

There are many types of agents depending on the roles they may fulfill. For instance, middle agents may act like intermediaries or brokers among systems. They support the flow of information by assisting in locating and connecting the information providers with the information requesters. In other words, they assist in the discovery of ontology models and services based upon a given description. A number of different types of middle agents have shown usefulness in the development of complex distributed multi-agent systems (Murry 1995). Agents may advertise their capabilities with a middle agent, and the method used for discovering and interacting with a service provider may vary depending on the type of middle agent used (cf: Dean et al. 2005). One example is that there may be a middle agent who mediates between requesters and providers by querying services whose advertisements match a requester's service query. The resulting messages are then sent from the provider to the requester via the middle agent.

This contrasts with a matchmaker agent, in which matchmakers do not participate in the agent-to-agent communication process directly; but rather, they match service requests with advertisements, and return these matches to the requesters. In these systems, matchmakers, or sometimes called yellow page agents, process advertisements, and blackboard agents collect requests, and broker agents coordinate both processes. The matchmaker is thus an information agent that helps make connections between various agents that request services and the agents that provide services.

1.4 Looking Forward

We have only begun to introduce key concepts related to our topic, but we needed to introduce some definitions before we delve into more complicated concepts. As people increasingly interact virtually in greater variety and with an expanding set of modalities, we are seeing a concomitant mimicry among the systems people use in the form of socially and biologically inspired technologies, which we will explore further in this book. We are also seeing the emergence of the blending of actualized and virtualized worlds such as in the form of augmented reality. This futuristic journey began with the idea that systems may inherently contain "meaningful" constructs that may eventually be entirely understood by a synthetic system. To date, markup languages can be combined with object-oriented features, which can implement expert capabilities and help us migrate from our current closed-systems approach to computing into an organic, open-system mode. It will be some time yet before many of these capabilities make it into the marketplace and become widely adopted. But as we shall see in the chapters to follow, an inkling of some of these characteristics has already been deployed, and those remaining are following closely on.

References

Berners-Lee, T., Hendler, J., & Lassila, O. (2001). The semantic web. Scientific American, May 17. http://www.cs.umd.edu/~golbeck/LBSC690/SemanticWeb.html. Accessed 14 April 2015.

Bray, T. (2003). On semantics and markup. Tbray.org. 4/14/2015: https://www.tbray.org/ongoing/When/200x/2003/04/09/SemanticMarkup. Accessed 14 April 2015.

Dean, M., et al. (2005). A semantic web services architecture. The Semantic Web Services Initiative Architecture Committee (SWSA). http://www.ai.sri.com/daml/services/swsa/note/swsa-note_v5.html.

Fresse, E., & Nexis, L. (2002). Using DAML+OIL as a constraint language for topic maps. XML Conference and Exposition 2002: IDEAlliance, pp. 2–5.

Miller, E. (1998). An introduction to resource description framework. http://dlib.org/dlib/may98/miller/05miller.html.

Murray, W. S., & Forster, K. I. (1995). Serial mechanisms in lexical access: The rank hypothesis. *Psychological Review, 111*(3), 721–756.

Lassila, O., & Swick, R. R. (1999). Resource description framework (RDF) model and syntax specification. W3C REC-rdf-syntax-19990222. http://www.w3.org/TR/1999/REC-rdf-syntax-19990222/. Accessed 18 April 2015..

Skyttner, L. (1996). *General systems theory*. London: MacMillan.

Workman, M., Ford, R., & Allen, W. (2008). A structuration agency approach to security policy enforcement in mobile ad hoc networks. *Information Security Journal, 17*, 267–277.

Chapter 2
Semantic Cognition and the Ontological to Epistemic Transformation: Using Technologies to Facilitate Understanding

Michael Workman and Daniel Riding

2.1 Introduction

In this chapter, we present the term "semantic cognition" as a way of introducing semantic systems. Semantic cognition involves the study of top–down, global, and unifying theories that explain observed social cognitive phenomena consistent with known bottom–up neurobiological processes of perception, memory, and language. It forms a foundation for explaining why some technologies work well and others do not. For instance, the problem of information, or cognitive, overload has become all too familiar. For example, cognitive overload can create unneeded stress and hurdles to effective decision-making in the workplace, thus hindering productivity (Adams 2007). Technologies have become quite good in terms of gathering and providing information to human consumers, but they have tended to worsen the information overload problem depending on their construction and use.

The development of technologies informed by semantic cognition emphasizes manipulating form to fit the task and function in terms of the design, development, and implementation, and in the evaluation of technologies relative to goal-oriented outcomes. Form to fit has many implications for how systems will be developed and utilized in the near future to improve human performance.

Structure, Structuration, and Agency

Agency in a structuration sense is anyone who acts within the formalized social structure of an organization. Thus, our use of the term "agency" represents individual behaviors that operate within a broad network of socio-structural influences

M. Workman (✉)
Advanced Research and Development, Security Policy Institute, Melbourne, FL, USA
e-mail: workmanfit@yahoo.com

D. Riding
Eastern Florida State College, Melbourne, USA

M. Workman (ed.), *Semantic Web*, DOI 10.1007/978-3-319-16658-2_2

(Chomsky 1996) that can be within or outside the formally defined organizational structures. Bandura et al. (1977) defined this triadic phenomenon as "agentic transactions, where people are producers as well as products of social systems" (p. 1). Agency exists then on three levels: direct personal agency (an individual's actions), proxy agency, which relies on others to act on one's behalf to secure individually desired outcomes, and collective agency, which is exercised through socially coordinative and interdependent effort (Bandura et al. 1977; Chomsky 1996). The notion of agency from this perspective contrasts with nondeterministic (chaotic) and nonrational "natural" processes that create the environments in which people operate either formally or informally (Table 2.1).

The reciprocal relationships between agentic action and social structures are referred to as the "duality of structure" by Giddens (1984). In these terms, structure is defined by the regularity of actions or patterns of behavior in collective social action, which become institutionalized. Agency is the human ability to make rational choices, and to affect others with consequential actions based on those choices that may coincide with or run counter to institutionalized structures.

Structuration, on the other hand, is a dynamic activity that emerges from social interaction. Particularly, social action relies on social structures, and social structures are created by means of social action. The existence of each and the interdependence of social action and social structures can thus be thought of as a constantly evolving dynamic. Thus, structures derive the rules and resources that enable form and substance in social life, but the structures themselves are neither form nor substance. Instead, they exist only in and through the activities of human agents. For example, people use language for communication with one another, and language is defined by the rules and protocols that objectify the concepts that people convey to each other (Chomsky 1996). The syntax structure of language is the arrangement of words in a sentence, and by their relationships of one to another (e.g., subject–predicate noun–verb phrase). The sentence structure establishes a well-defined grammar that people use to communicate.

However, language is also generative and productive and an inherently novel activity, allowing people to create sentences using the syntax rather than to simply memorize and repeat them (Chomsky 1979). In similar fashion, institutionalized

Table 2.1 Agentic attributes

Autonomy	The ability to pursue an individual set of goals and make decisions by monitoring events and changes within one's environment
Proactivity	The ability to take action and make requests of other agents based on one's own set of goals
Reactivity	The ability to take requests from other agents and react to and evaluate external events and adapt one's behavior and make reflexive decisions to carry out the tasks toward goal achievement
Social cooperation	The ability to behave socially, to interact and communicate with other agents
Negotiation	The ability to conduct organized conversations to achieve a degree of cooperation with other agents
Adaptation	The ability to improve performance over time when interacting with the environment in which an agent is embedded

structures regulate agentic behavior, but agents may also disrupt institutionalized structures. The defining features within structuration theory that explain how these processes work are: signification, legitimation, and domination. Signification concerns how meaning is cocreated and interpreted by agents, legitimation encompasses the norms and rules for acceptable behavior, and domination refers to power, influence, and control over resources (Giddens 1984). Collectively, the signification, legitimation, and domination constitute the institutionalized structuration processes. Agentic interaction with these processes creates the communicative structure, authoritative structure, and allocative structure, respectively.

It is important to note that agency behaviors can be modeled in adaptive information systems to act more like human beings so that they can be more compatible with how human beings work and solve problems. For example, modeling these sociobiological artifacts in software have led to the development of epigenetic systems (Bjorklund 1995) in which linear models have become supplanted by more dynamically organized computational models that perform multiple operations simultaneously and interactively with the environment in which it operates (Bandura et al. 1977). The software, or machine, is thus evolving and operating by learning from its environment in an open-ended fashion. Thus, epigenesis from a sociobiological perspective asserts that new structures and functions emerge during the course of developmental interaction between all levels of the agentic biological and environmental conditions (Bjorklund 1995). The notion of agency from this perspective contrasts with nondeterministic (chaotic) and nonrational "natural" processes that create the environments where people are embedded (Beck et al. 1994).

Agency and Agent Systems

Big data analytics draw from mining patterns out of data warehouses or distributed stores. This is a closed system, that is, information is pulled out of an environment and stored away in a large database where it is later examined for patterns by using various analytics. Much may have changed in the dynamic environment since the time the data were extrapolated into the closed system. The closed-system static model of pattern discovery is inherently limited. Moreover, with data warehousing analytics, the user must provide the problem context. By way of using the Web as an analogy, a user must "drive" the search for information with the assistance of a technology such as a crawler or bot. This has widely recognized limitations.

The Web is filled with a sea of electronic texts and images. When you look for something of interest, unless someone provides you with a URL link where you can find relevant material, you will then have to resort to a search engine that gathers up links to everything that it thinks is related to my topic. It is then necessary for you to begin an extensive hunt, sifting through the links looking for possibilities. When you find a page that sounds interesting and begin reading through the material, you will likely discover that it is not what you had in mind. Many of the pages in the Web are cluttered with a multiplicity of subjects, and they are littered with links tempting you to divert your limited attention to another realm, causing you to abandon the original quest in favor of a newly piqued interest (Palmer 2001). Because of their agentic and social attributes, agent-based systems have the potential to help alleviate some of these problems by seeking goals and making evaluations. For example, you may be working in an office in Florida when your boss calls and asks you to attend a meeting with a customer in Dallas to present the company's technical strategy. You then give instructions to an agent to gather intelligence on the customer so that you can frame the presentation for the audience. You may instruct him/her to find published strategies with which to compare so that the customer will see that you have come prepared, and you may also instruct the agent to book the trip, finding the best plane fares for the flights you would want to take, and a hotel near the customer site. To perform these functions, the agent cooperates with other agents (in multi-agent systems, or MAS) to exchange information, resources, and tasks.

The ecosystem in which agents operate is organic. The systems generate descriptions of things and events in the system (called models) and the rules (also in the form of a model) for other agents to use when operating on these models. The systems are not only self-describing, but because the models are dynamically generated within the ecosystem, they are self-defining. Furthermore, models may be advertised and discovered by agents. An agent may even traverse the places where models are advertised and "look" for things and do things. Such a system would be self-renewing, because it can import and export resources. Self-defining, self-renewing, and self-organizing characteristics define an organic system (Bertalanffy 1968).

Goal-Directed Agents

Many systems such as found in many contemporary network and application monitoring use simple stationary agents. In a semantic sense, agents take on more complex behaviors including mobility (Usbeck and Beal 2011). From a semantic perspective, a software agent is an "independent software program with real-time decision-making abilities that acts intelligently and autonomously to deliver useful services" (Agentis Software 2008, p. 1). Goal-directed agents are a special case (cf. BBN Technologies 2004). These agent frameworks are able to adapt in dynamic environments by allowing them to deviate from predefined plans according to their situational awareness (Fig. 2.1).

Goal-directed agents perform a series of steps to carry out a plan, while the agent monitors its environment for substantive changes relative to achieving its defined goals. An agent may choose a different plan, set new goals, and update its "understanding" if it encounters impediments. This ability to "infer" based on changes in the ecosystem is what distinguishes goal-directed agents from their more static predecessors. With goal-directed agents, new plans may be added without affecting the existing plans because plans are independent of one another. Moreover, because agents assemble their execution contexts at run-time, execution paths and error recovery procedures are not required during their design and development (Agentis Software 2008).

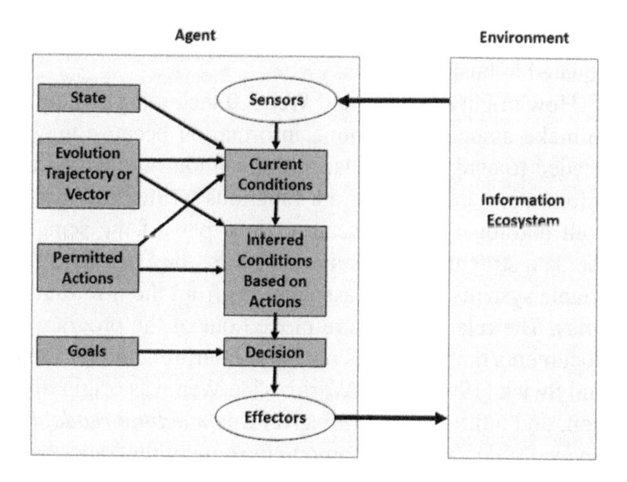

Fig. 2.1 Goal agent architecture

The Problem of Meaning

Building semantic systems stems from human cognition and perception. Thus, a discussion is warranted here to explain what follows.

At the heart of semantic systems is the definition and derivation of meaning. Even with the promise of these more organic technologies, "meaning" is a human construction. If you say, tear, what does this word mean? To answer this question, we need to know the relationship of this word to other words. We need it in context. The word means something different if you say: You have a tear in your shirt versus you have a tear in your eye. One word with two meanings is one level of the semantic problem.

We also have the inverse—many words with one meaning. The antonym and synonym issues are still only half the semantic picture. There are other problems we put in the category of transformational grammar (Chomsky,1979). A door may be opened, or it may be open. We also have the issue of some words operating as verbs in one context and nouns in another—wave, for example—look at the wave versus wave at the crowd. To begin to address this problem, we need some way to describe an entity. The first part of the semantic problem deals with the antonym and synonym problem, and hence the relationships between things are important. However, it is not as simple as that, attributes that define objects can be different, such as with the following:

Teacher: Teacher Name, College, Discipline.
Teacher: Teacher ID, Teaching Philosophy, Degree

The two entities called Teacher consist of different attributes. Some of these attributes may be the same, such as Discipline and Philosophy, but maybe not. A real example is found at the Coca-Cola Corporation where they use independent distributors and bottlers worldwide. Not only do these entities use different languages but also each has different notions of entities as defined by their attributes.

A customer in Bulgaria is not the same as a customer in Montreal. They cannot be equated in business terms.

How might semantic and Web 3.0 technologies help? Most systems are not able to make associations among information because they do not have the structures needed to analyze the relationships among the data; they are only able to process information and perform the functions written into programmatic logic. However, with ontologies, the structures carry part of the semantic association inherent in the data structures themselves. That is, they provide relationships among data that enable systems to make associations from the information based on predetermined rules. The relationships are moved out of the program code and placed inside the documents that programs read and interpret, and reason over. According to Lassila and Swick (1999), "The World Wide Web was originally built for human consumption, and although everything on it is *machine-readable,* this data is not *machine-understandable.*" This is among the core of the issues being resolved. To understand how, we need to present an overview of human cognition.

2.2 Cognition Overview

There are (at least) two schools of thought on memory processing and cognition—the Information Processing Approach, where an analogy between the mind and a digital computer is made, and the Ecological Perspective, where the focus is on the dynamics of the environment a person is in, including with machines and people. Informational Processing is conceptualized as where the mind is more "computational" using memory to access memory and form a representation with meaning to a stimulus, whereas the Ecological Perspective relies on a person's perception of the environment around them and their actions form the basis of the conscious mind. The information processing perspective is based on mind–environment dualism, while the ecological perspective is based on mind–environment duality (Cooke et al. 2004).

Memory and Cognition

It is widely recognized that while the capacity of long-term memory is, in theory, virtually unbounded, attentional or working memory is severely limited (Halford et al. 2005). Nevertheless, human brains have the ability to process some kinds of information in simultaneous and nonlinear ways. For example, one may be deeply engrossed in a conversation with her friend and suddenly feel a spider crawling on her hand. Her sensory systems perceive the tiny legs of the spider on her skin and alert her attention; her hypothalamus releases neurochemicals that elicits a fear response to the potential spider bite, she sweeps the spider from her hand and continues her conversation. The person in this situation reacts unconsciously before her schematic knowledge structure stored in memory has processed the behavior (Gioia and Poole 1984). This type of "multiprocessing" indicates that working memory acts as an event receiver, where stimuli compete for "time slices" of attention (Anderson 1983, 2000).

To highlight this point, Dennett (1997) presented the multiple drafts of consciousness theory in which he posited that our conceptions and perceptions of reality are formed in working memory by receiving "snapshots" of the activities processed in different parts of our brains. "Pasting" these snapshots together is somewhat analogous to how photo frames are strung together to make motion pictures (movies). Interestingly, these "realities" are not as contiguous as they might seem in a movie; instead, they are more akin to showing chunks of several different movies in an alternating fashion. However, this does not mean that our attention oscillates between different static frames of apprehension as I might have implied with the simple movie analogy—rather, our brains process information and stimuli with varying degrees of conscious attention in a very fluid and dynamic fashion (Anderson 2000; Bargh and Morsella 2008).

Examining these features reveals the notions of implicit and explicit cognition (Hutchins et al. 2013). Implicit cognition is defined as those processes which are automatic, effortless (in terms of working memory), unconscious, and involuntary, whereas explicit cognition is defined as the intentional use of working memory (Schacter 1995). Given these distinctions, we may also consider "thought" as a memory retrieval process, whereas "thinking" is a creative reconstruction from what has been learned or experienced, or as a process of imagination or concentration (Jensen et al. 1997).

While many functions are specifically performed in well-defined parts of the brain, such as speech and language (most often located in the left hemisphere called Wernicke's and Broca's areas), many portions of the brain are malleable insofar as they "rewire" neural connections, a property known as plasticity. It is intriguing to note that owing to neuroplasticity, the more one attends to a particular stimulus, generally the more readily one comes to recognize or focus on it (Bransford and Johnson 1972). One reason for this is because more frequently used neural pathways are more readily primed, along with their associated cognitive schema (Barnhardt 2005). As Bargh and Morsella (2008) noted, "cognition research on priming and automaticity effects have shown the existence of sophisticated, flexible, and adaptive unconscious behavior guidance systems" (p. 78).

Priming effects and automaticity may be crudely thought of as water following the paths of least resistance—in other words, neural pathways that have been recently or intensively utilized are more easily charged or activated (Craik and Lockhart 1972; Khemlani and Johnson-Laird 2013). Cognitive schema may be thought of as a network of concepts, rules, and protocols (McNally et al. 2001). To illustrate, a procedural schema for ordering food when primed with the word "restaurant" causes people to retrieve a specific set of expectations for their prototype of the restaurant concept. When a prime is modified, such as in the phrase "fast-food restaurant," the schema is also modified (Schacter 1995).

Nevertheless, despite this cognitive flexibility (Shabata and Omura 2012), people tend to lean either toward implicit or explicit cognitive dominance, especially when under time pressure to solve complicated or subjective problems (Barnhardt 2005; Gawronski and Bodenhausen 2006; Richardson-Klavehn et al. 2002). Moreover, since working memory is limited, people develop "habits" because they are cognitively economical (Halford et al. 2005). According to Biel and Dahlstrand (2005), habits derive from deeply embedded and richly encoded thoughts and behaviors built up over time, whereas explicit cognition (including the use of newly learned problem-solving strategies or principles used for making judgment calls) must be remembered and intentionally used.

Moving from the concepts of implicit and explicit cognition, we look at metacognition, which is "knowing what you know." In other words, metacognitive processes create awareness and help coordinate cognition involved in acquiring perceptual, conceptual, and thinking feedback, and monitoring progress toward task solutions (Sternberg 1977). Improving metacognition enables individuals to be better equipped to attend to and interpret relevant information, and use this information to decide how to act and perform effectively (Blume et al. 2013; Engonopoulos et al. 2013). The utilization of metacognitive strategy is also a key difference between expert and novice learners, where the expert learner plans cognitive strategies, monitors them, and will revise strategies to meet goals (Goldstein and Ford 2002). This use of metacognitive strategizing can be useful when dealing with information overload.

Next, when people are inundated with information or when information becomes extremely complex, they experience cognitive information overload (Killmer and Koppel 2002; Watson and Tharp 2013). Since durable information is stored in the form of organized schemata in long-term memory, semantically enriched information helps free up working memory resources and hence allows the limited capacity of working memory and explicit cognition to address anomalies or attend to the more novel features in the information conveyed, and permit cognitive processes to operate that otherwise would overburden working memory (Hutchins et al. 2013; Paas et al. 2003; Seitz 2013; Shabata and Omura 2012). There are emotional and physiological reactivity effects associated with subjective job overload of workers leading to burnout caused by demands made upon them in the work environment and the resources available to them (Shirom 2003).

Information Structure and Semantics

Consider that the bulk of the information with which we are presented and utilize comes to us in a linear form, such as lines on a page that you are reading. One can imagine that this does not capitalize on the brain's natural ability to process information in simultaneous and nonlinear ways. As an example of this linearity, if we asked the question, "What does the word tear mean?" It is unlikely that someone would not be able to tell unless we stated that you have "a tear in your eye" versus you have "a tear in your shirt." We rely on this dependable and relational information structure so that we can "make sense."

In prose, our ability to gain and share knowledge in this way can be described as transformational grammar (Chomsky 1979). Vocabulary rules help to convey semantics because they determine the objective measures by which people draw conclusions and make inferences about ideas. Transformation grammar involves two levels—a deep structure and a surface structure. The deep structure is essentially that of meaning (or intended meaning) encoded with the surface structure, which is that of syntax. To formulate a conception of meaning, or to draw conclusions and make inferences about an intended meaning, the rules and relationships among the words or concepts must be known (Shiffrin and Schneider 1977). Transformational grammar, therefore, is the system of rules and relationships that transform ideas from one structural level to another (Kozma 1991; Trafton and Trickett 2001).

Beyond sense making from information structure, another important aspect of semantic cognition is situational awareness. Endsley et al. (2003) described situational awareness as cognition on three levels: (1) comprehending or perceiving relevant elements in a situation, (2) understanding the meaning of the elements, and (3) the application of the understanding such as to be able to project future states and make inferences. Consequently, situational awareness is a type of "cognitive map" that people develop as they receive information.

While information may be received in many forms (e.g., sound or touch), the majority of information with which people presently work is visual (Card et al. 1999). We have concentrated on visual information so far because this has been the dominant form of information representation in business to date, especially that of written texts. At this juncture, however, we note that visual information has other conveyances, such as with images and drawings. If we consider how these are perceived by our visual sensory systems, and our apprehension of meaning, we might take, for example, a painting we appreciate. The painting conveys information to us in a holistic and simultaneous manner (Langer 1957), but it may leave us with a vague subjective impression of what the painter intended with his or her rendering and what we determined it to mean. The reason why we may not be able to objectively interpret the meaning of the painting is that it lacks transformational grammar. Although some experimentation has been done using graphical linguistics, cuneiforms, symbols, and various forms of isotypes (cf. Lidwell et al. 2003), there has yet to be a consolidation in terms of principles that could enable generalized and objective interpretations.

Indeed, despite a large stream of cognitive and neuroscience theory and literature on visual perception, attention, memory, and linguistics, this is one area where human factors research has traditionally lagged behind the underlying work related to information storage and retrieval theory (Gavrilova and Voinov 2007). This is an interesting issue because underlying storage and retrieval research (cf. McBride 2004) have been utilizing semantic and cognitive theory to drive the development of markup such as RDF and OWL for more than two decades.

The disparity between the semantically rich underlying description logics and the representation of the information models in visual displays of information begs for more theory-driven implementations based on semantic cognition. An interesting feature of these description logics is that the typical linear or hierarchical data structures, such as found in XML, are supplanted by richer embedded relational structures (as in the case of RDF) with some of the rule features of transformational grammar (as in the case of OWL). These are interesting features because the data structures more closely resemble cognitive schema, inherently.

2.3 Visual Perception

In the previous section, we presented an overview of cognition and introduced that the ways information can be structured may either aid or complicate the extraction of meaning from the available data or information. We discussed linear representations of information, and presented that the rules and relationships among components enable our understanding. The purpose of this section is to describe the processes of visual perception and visual perceptual memory in order to understand the human factors and visual communication issues involved in effective design, implementation, and utilization of semantic technologies that best facilitate visual apprehension of information rendered by them. We take a quick physiology tour of vision before we get into some practical points with implications for semantic information display especially of high-density data that are time sensitive.

Vision and Visual Perception

Visual perception is the way in which we interpret the information gathered (and processed) by the eyes. We first sense the presence of a visual stimulus, and we perceive what it is. Light waves enter the eye where images are inverted by the lens. The light waves are then projected onto the reactive surface of the eye, called the retina. The retinal surface consists of three layers of neurons: rods and cones, bipolar cells, and ganglion cells. The rods and cones form the back layer of the eye, and these are the neurons that are first stimulated by the light. Only a fraction of the original light energy is registered on the retina, the rest is absorbed and scattered by the fluid and structures within the eye.

Patterns of neural firing from the rods and cones are forwarded to the bipolar cells, which collect the messages and pass them along to the ganglion cells. These have extended axons (stems) that converge at the rear of the eye in the optic nerve. The image that is captured in the eye has a blind spot (called a scotoma) which is "filled in" by other cognitive and perceptual processes. The optic nerve exits the eye and continues onto the visual cortex in the occipital lobe in the back of the brain, first crossing midbrain between the left and right hemispheres such that the left eye's image is projected to the right side in the occipital lobe and vice versa. The image that is projected to the back of the brain is upside down, and must be set upright by our mental processes. By the time these processes transform, analyze, and summarize the visual input, the message that finally reaches the visual cortex represents preprocessed and compressed record of the original visual image.

When we read, we get the sense that our eyes consume visual information in a continuous fashion. However, the eye actually sweeps from one point to another in movements called saccades, and then pauses or fixates, while the eye encodes the visual information. While this eye movement is fairly rapid (about 100 ms), it takes about twice that long to trigger the movement itself. During the saccade, there is suppression of the normal visual processes, and for the most part, the eye only takes in visual information during the fixation period, roughly 20 ms (Thomas and Irwin 2006). This means that there is enough time for about three complete visual cycles of fixation-then-saccade per second. Each cycle of the process registers a distinct and separate visual image, although, generally speaking, the scenes are fairly similar and only a radical shift in gaze would make one input completely different from the previous one (Urakawa et al. 2010).

Studies (e.g., Haber 1983) show that there are differences in visual perception between viewing a natural environment versus a computer screen. For one thing, the focus of our field of vision is narrower when working with a computer screen than when we are attending to visual stimuli in our natural environment. Also, there are differences in what is called dimensionality.

No matter how an image may appear on a computer screen (even if rendered in three dimensional—3D), the screen can only display on a flat surface. Another important characteristic is that visual information is only briefly stored in memory. The duration of time that an image persists in memory beyond its physical duration depends on the complexity of the information that is absorbed during the encoding process.

Visual Memory Processing

Just as we have the ability to remember a phone number, our visual sensory system has the ability to persist information. There are also individual differences in relation to three types of intelligence posited by Carroll (1993). Visual spatial reasoning is a type of intelligence measure of one's ability to see foreground, back, distance, and speed. In terms of computer image processing, Averbach and Sperling (1961) performed a series of interesting experiments that showed, on average, people have deterioration in visual recollection as the information complexity increases. For example, when up to four items were presented in their studies, subjects' recollection was nearly complete, but when up to 12 items were presented, recollection deteriorated to only a 37% level of accuracy. Furthermore, they found that this poor level of accuracy remained essentially the same even for exposures of the visual stimuli lasting for a long time—in a visual sense, as long as 500 ms.

Consequently, in general, people have a span of visual apprehension consisting of approximately four or five items presented simultaneously, although there is some variability relative to visual image persistence based on the contrast and background upon which images were rendered (Greene 2007; Irwin and Thomas 2008). When dark fields were presented before and after a visual stimulus (consisting of

18 letters), visual memory was enhanced (just as a lightning bolt is more visible in a nighttime storm than a daytime storm because of the contrast illumination). More specifically, Averbach and Sperling (1961) found that more than 50 % of the letters presented were recalled well after a 2-s delay when dark fields were used, but in contrast, accuracy dropped to 50 % after only a quarter of a second when light fields were used.

Research (e.g., Logan and Cowan 1984) has shown that a later visual stimulus can drastically affect the perception of an earlier one. This effect is called backward masking. The masking stimulus, if it occurs soon enough after the display of the original, interferes with the perception of the earlier stimulus presented at the same position. In some studies of backward masking (e.g., Becker et al. 2000), subjects literally claim that they see only the subsequent (masked) stimulus even though their other performance indicators (such as timings in recognitions of previously seen versus unseen items) suggest that the sensory system did indeed register the first stimulus (Irwin and Thomas 2008; Thomas and Irwin 2006). When the contents of visual sensory memory are degraded by subsequent visual stimuli, the loss of the original information is called erasure. Studies (Shabata and Omura 2012; also see the seminal work: Stroop 1935, and long lineage of confirmatory work, e.g., Sternberg 1977) have shown that the information stored in working memory of visual images may not simply erase or distract from information that was recently presented, but rather can facilitate the anticipation of information that might appear next, called proactive interference.

A final point for our purposes in this section is that visual information processing involves feature detection. A feature is a simple pattern or fragment or component that can appear in combination with other features across a wide range of stimulus patterns. Studies (Neisser 1967) of unelaborated features (those without surrounding context) suggest that we read printed text by first extracting individual features from the patterns, then combining the features into recognizable letters, and then combining the letters to identify the words. With surrounding context, we use the cognitive heuristic of "likeness" to infer correct from misspelled words if there is enough context from which to make the inference. As an example, most people are unable to see that a single word is misspelled such that "slevin" might be unintelligible, unless we write that "four score and slevin years ago." The influences of surrounding information along with a person's own previous knowledge are critically important to understanding visual information.

2.4 Memory and Attention

In the previous section, we spent some time on visual perception because this is an exciting new area of applied research and development that semantic technologies will soon leverage. Structuring visual data, especially high-density data, has been and will continue to be one of the major bottlenecks to comprehension of complex data and situational awareness. Many human factors experts (along with popular bloggers, display design consultants, and other laypeople) have tried to show how to design cognitively economical data displays, but most have failed because they merely take one display media, such as a pie chart, and show it in a different form without any understanding of human vision or semantic cognition. Semantic technologies are poised to be a game changer in that regard. In this section, we devote some time to cognitive processes, and how working and long-term memory work, relative to attention.

It is interesting that we seem to be able to do so many things at once, but that we can only seem to concentrate on a single thing at a time, and still some of our behaviors are habitual and operate without any conscious attention at all (Halford et al. 2005). We perceive our attention as a central controller of sorts (cf. Anderson 1983, 2000), and that it directs all of our other cognitive activities, but this is an illusion (Brünken et al. 2002). Working memory and attention act more like event processors—they respond to stimuli that stream in from all of our perceptual–sensory and mental activities going on in various parts or our brains (Bargh and Morsella 2008; Barnhardt 2005; Hazeltine et al. 2006). In this section, we briefly cover some of the underpinning theories that explain our "attentional" cognition and draw some implications for the design and implementation of technologies as these relate to human performance.

Working Memory

Colloquially (because the term was applied in early research), we often refer to our *short-term memory*. The term "short-term memory" is no longer used because we have come to understand that this cognitive activity is much more than a simple storage system (Baddeley and Hitch 1974). The term working memory has replaced short-term memory in the academic lexicon because the cognitive function it comprises involves information, concepts, percepts, and operations. Researchers originally thought that short-term memory consisted of only that which we are consciously aware. More modern research into metacognition (the term applied to knowing what we know) has indicated that some mental processes occur in working memory that are not revealed to consciousness—and, in fact, they are performed automatically (Brünken et al. 2002). People may be aware of the contents of their working memory, but they are not necessarily aware of the processes that occur to retrieve information and operate on them (Zaccaro et al. 2001).

The early work of Miller (1967) showed that people are able to store an average of seven items (plus or minus two) in working memory. Subsequent research (Cowan 2000; Halford et al. 2005) has shown that this greatly depends on the complexity of the information and the number of operations that people perform at a given time—and that the number is somewhere below seven items. However, techniques such as "chunking" information into groups (e.g., a phone number may be chunked as a single item), and creating associations between concepts (such as by using rhymes or peg words) can augment this limited faculty.

Augmenting is often called recoding—because groups of concepts are used to form an associative network, which makes newly formed units more "meaningful" and hence, at least partially, automatic—akin to grabbing a link in a chain and pulling on it brings other links with it. Thus, working memory as explained is different from long-term memory, and this has been observed physiologically (Bargh and Morsella 2008; Barnhardt 2005). Working memory involves anatomical components deep in the central part of the brain, whereas long-term memories are stored in the outer cortex. The interesting aspect of this, however, is that cognitive functions "blur" because cognition is a symphony, but there is nothing analogous to a conductor except by way of the term attention.

One way to imagine this is via an interesting phenomenon called the serial position effect in which information that is first seen (primacy) or is most recently presented (recency) is most readily recalled. We see in the serial position effect the interactions among working memory, attention, and long-term memory. It is also interesting to realize that memories are handled differently by the brain depending on whether they are learned skills and procedures, or experienced, or derived as a cognitive process such as performing calculations, or processes such as doing mental image rotations.

Types of Memory

Based on what we have covered thus far, we might consider that there are "types" of memory processes that we may label as *procedural, episodic, semantic,* and *declarative* memory. Procedural memory involves "how to do something" such as the processes invoked when we, for example, drive an automobile, or balance our checkbooks. Episodic memory is autobiographical, that is, a memory of personal experiences, and semantic memory, which refers to a recollection about meaningful events or our "world knowledge." Declarative memory (representations of learned knowledge) includes both semantic and episodic memory components (Schacter et al. 2011; Tulving 1972).

As indicated earlier, our experiences are wrapped up with the tasks and skills that we learn and are stored away for future reference in a relational manner in what is called a cognitive script or schema. Thus, our memory system stores separate groups of information together to the extent that those separate groups are related to each other. To understand new experiences, we use what we already know in a conceptually driven fashion. We call these conceptual groupings schemata, or general world knowledge. Those new experiences then become part of our elaborated knowledge structures, and continue to assist in later cognitive cycles of conceptually driven processing. These integrative memory tendencies have a feature known as encoding specificity, which refers to a phenomenon that when people learn a task or other information, they also encode (integrate and store) information about their surroundings. For example, students perform better on a test when they take the test in the same classroom in which they learned the materials rather than if they take the test in a different classroom.

Information is more durably stored and more readily recalled from long-term memory when the information is semantically enriched compared to when those items were learned in a less meaningful way. Elaboration is a technique to semantically enrich information by adding context to concepts. For example, in a series of experiments, Morris et al. (1999) showed that the concept "bear" was recalled 86 % of the time when it was memorized in a sentence such as, "The bear ran through the woods," compared to 63 % of the time when "bear" was memorized in a list of rhyming words such as "bear, hair, care," and fell to only 33 % when "bear" was memorized merely as part of a random list of words. This shows that elaborative rehearsal occurs when people do not merely read text, but rather search for connections and relationships that make the text more memorable. In addition, retention of information persists longer when given certain features that make the words in a list more distinctive, such as underlined, in a different font, or a different color.

Cognitive Processing

Conscious information processes are open to our awareness and occur only with our intention; that is, they are deliberately performed. Because of this, conscious processes require and consume some of our available resources in working memory. On the other hand, automaticity is the property that some cognitive functions have to operate automatically, and is central to how attention, pattern recognition, and memory work.

Three characteristics define an automatic process as indicated earlier: (1) the process occurs without a person's intention, (2) the process is not revealed to consciousness (attention), and (3) it does not consume working memory ("attentional") resources (Schacter 1995). A clever mechanism to discover how all our cognitive processes function together in concert is through what is called a dual-task test. A dual-task test divides conscious from automatic processes by giving the research subject a primary (intentional) task to perform, and then measuring the time it takes for him or her to react to a secondary cue or task (an automatic response). In one form of this type of test, a subject is given a timed writing task to perform, while simultaneously listening to a narrative on headphones. Subjects are able to perform the writing task virtually as fast as if they had no auditory accompaniment—filtering out the auditory narrative. However, when the subject's name is said in the narrative, it interrupts his or her performance. This indicates that there are cognitive processes, operating below the level of consciousness, that are attending to sounds in one's environment. This is a specific example of our ability to attend to information and cues on levels below that of consciousness.

Illustrating this concept with visual information, Stroop's (1935) seminal research showed words to subjects such as "RED," "GREEN," "BLUE," and "YELLOW" in colors other than the words (e.g., the word "RED" might be written in blue ink). Subjects were required to name the ink color rather than the printed words as quickly as possible. Stroop found significant timing delays (interference) when the color name and ink color were different. This indicates that accessing the meaning of the written symbols, such as RED, is automatic—it happens whether subjects wanted it to or not. We refer to this effect as "priming"—a word automatically activates a given meaning in memory, and as a consequence, primes or activates meanings closely associated with it. Priming makes related meanings easier and faster to retrieve because the information is cognitively networked together by association. Automatic processes, such as priming, do not consume working memory resources, so then why would priming interfere with conscious functioning relative to the Stroop test? Consciously naming ink colors is interrupted because while cognitive access to the meaning of the word is performed automatically, the interference occurs at a later stage in our cognitive processing.

The Stroop test shows that interference is created because of the incompatibility of responses that are competing simultaneously for working memory resources: saying the ink color word when the automatic reading process has primed a different color word. We find no such interference when different ink colors are used to print words such as "PEN," "TABLE," "CHAIR," "COMPUTER," and so on. The automatic and conscious processes interfere with each other only when they compete for the same cognitive response mechanism. Thus, we can see that relationships we form among concepts, even if processed in different parts of the brain, are crucially important to human performance because of the brain's integrative tendencies.

As we mentioned before, thinking is a creative process, whereas having a thought is basically a memory retrieval process. Underpinning the theories of attention and cognitive processing is the idea of implicit and explicit cognition. Implicit cognition results from automatic cognitive processes, as we have stated. Recall that automatic cognitive processes are effortless, unconscious, and involuntary. It is rarely the case, however, for all three of these features to hold simultaneously, but it should be pointed out that *ballisticity* (Logan and Cowan 1984), a feature of a cognitive process to run to completion once started without the need of conscious monitoring, is common to all implicit processes (Bargh and Morsella 2008).

Explicit cognition results from intentional processing that are effortful and conscious (Jacoby 1991). Conscious monitoring in this context refers to the intentional setting of the goals of processing and intentional evaluation of its outputs. Thus, according to this conceptualization of cognition, a process is implicit if it (due to genetic "wiring" or routinization by practice) has acquired the ability to run without conscious monitoring, whereas intentional cognition requires conscious monitoring and relies on working memory (Richardson-Klavehn et al. 2002). Taking this into account, Baddeley and Hitch (1974) proposed a model of working memory consisting of a number of semi-independent memory subsystems that function implicitly, which are coordinated centrally by a limited capacity "executive" that functions explicitly. Their model suggests that there are separate stores for verbal and visual information; for example, a "visuospatial sketch pad" is responsible for temporary storage of visual–spatial information, with the central executive being responsible for coordinating and controlling this, and other peripheral subsystems (Barnhardt 2005). Their model also highlights the effects of explicit cognitive processing of information encoded serially. Human cognition works in this fashion essentially as a linear scanning system (Halford et al. 2005).

For instance, in an auditory channel, people use an "articulatory loop" to rehearse and elaborate on information they hear to form cognitive schema. In a visual channel, people make brief scans across the series of symbols and then fixate momentarily (saccades), while they encode the information into cognitive schema (Smith and Jonides 1995). These encoding processes consume working memory resources, and the effect on performance is a product of the available working memory resources. As information complexity increases, there is greater serialization of information increasing cognitive load, which drains cognitive resources, and task performance deteriorates (Hazeltine et al. 2006).

Next, Anderson's (2000) model of human cognitive architecture asserts that only the information to which one attends and processes through adequate elaborative rehearsal is spread to the long-term memory. Long-term memory can store schemata and subsequently retrieve them with varying degrees of automaticity (Reder and Schunn 1996). The capacity of long-term memory is, in theory, virtually unbounded, but people are not directly cognizant of their long-term memories until they retrieve the schema into their working memory, which is greatly limited—with seven concepts (plus or minus two) being the upper bound (Cowan 2000; Halford et al. 2005; Miller 1956).

Since durable information is stored in the form of organized schemata in long-term memory, rendering information effectively to people can free up working memory resources and hence allow the limited capacity of explicit ("attentional") cognition to address anomalies or attend to the more novel features in the information conveyed, and as these schemata allow for enriched encoding and more efficient information transfer and retrieval from the long-term memory, they allow cognitive processes to operate that otherwise would overburden working memory. From research (e.g., Lord and Maher 2002) into our understanding and processing of concepts—the essence of semantics—we find that a distinguishing feature of semantic memory is based on acquired knowledge about the relatedness of concepts. Episodic memory, in contrast, represents empirically acquired experience and later evaluated as we face new situations. This highlights the difference between remembered versus constructed meaning.

When already-known information influences our memory for new events, we call it conceptually driven processing. In addition, there are alternative models which have been proposed to deal with a more individual-centered method of information processing consisting of rational, limited capacity, expert, and cybernetic models (Lord and Maher 1990). The rational model assumes that individuals process all relevant information for an outcome. The limited capacity model explains the use of cognitive simplification methods, such as satisficing and the use of heuristics, to reach a decision. The expert model is when a person uses existing knowledge structures that are highly organized and developed of a content domain to supplement simplified processing. Finally, the cybernetic model proposes that the processing of information and actions associated with it are dynamic over time, and uses simple heuristics procedures, just as the limited capacity model does, but is affected by feedback.

Semantic Relatedness and Cognition

In order for us to make plans and decisions, we need to collect information and draw inferences about causes and effects. This function relies on the semantic aspects of cognition. From the previous subsections, we have gotten a sense of how semantic memory operates, and have gathered the importance of relationships among concepts in the construction, storage, and retrieval of meaning. From the earliest studies of memory, it has been shown that nonsense words are less memorable than meaningful words, and that contextual stories are more understandable than non-contextual ones. It is the association among concepts that gives them their meaning.

Collins and Quillian (1972) developed a model of how information is mentally grouped together into meaningful units as part of a network of concepts. They described the structure of semantic information as an "interrelated set of concepts, or related body of knowledge where each concept in the network is represented as a discrete element with associative pathways to other discrete elements." For example, a bird might be semantically linked with "flying," "feathers," and "eggs." We refer to the phenomenon spreading activation, which is the mental activity of assessing and retrieving information from this network. With spreading activation, retrieving one nodal concept leads to the activation of all the other interconnected nodes. The spreading of these nodes leads to almost simultaneous thinking of other mental representations (Fig. 2.2).

Research (Sternberg 1977) has shown that the network of semantic concepts also contain property and superordinate pathways, as well as ordering concepts in subject and predicate terms such that the less related the concepts (called semantic distance), the more difficult the retrieval of the information. For example, in experiments, "a robin is a bird" is retrieved more readily than is "a robin is an animal." Of course, these categorizations and their subsequent retrieval depend on learning, both as a process of being taught and from experience, such as "a bee stings."

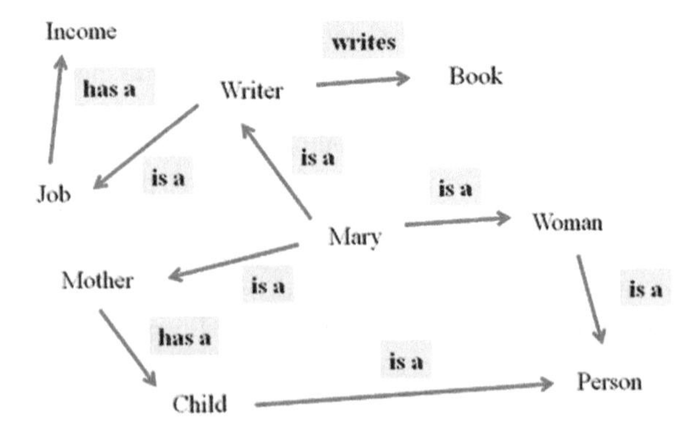

Fig. 2.2 Cognitive schema and semantic network

What we perceive from our visual sensory system is based on detecting "features." Earlier, we read that looking at information on a computer screen is different from seeing information in our natural environment. When we perceive a thing, we classify it, but we find that people perceive natural versus artificial categories differently. Artificial categorization tends to be more discrete, whereas natural categorization tends to be "fuzzier." This is a property of "structure"—artificial categories (a circle is round) are generally given more structural features than natural, real-world categories (the Earth is round, sort of). People formulate perceptual categories based on structure, and also based on analogous structures experienced in nature. For example, in one set of studies (Rosch 1975), subjects were asked to rate a list of category membership based on their representativeness or typicality for a given class of objects, such as: "which is a better member of the 'dog' category, a Collie or Poodle?" It was found that categorization used to classify a member of a set (e.g., a Poodle is a dog) depends on the rated typicality from the subject's own experience. This research shows that real-world category membership is not a collection of bundles of independent and objective features, nor are they classified into categories only because of the presence or absence of certain features (e.g., is it square? Is it shaded? Is it on the left?).

Instead, real-world categories indicate ill-defined and uncertain membership for a variety of instances: Is a "sled" a toy or a vehicle? Because concepts and categories that occur in the natural setting of our experience have a complex internal structure, "fitness" is important. That is, the category "dog" has an internal structure in which some members are better fits or are more representative than others. The fit comes from the frequency of association that people have, and hence differ depending on various factors such as geography and culture. For example, people in Mexico are more likely to associate Chihuahua with "dog" than Poodle, whereas in France, the opposite is true. We have a central meaning for each category and concept, and these can be represented, in a semantic distance sense, in terms of prototypical similarities and dissimilarities based on our experience.

Semantic Priming

Any stimulus that is presented first such that it leads to anticipation and hence influences some later process is called a *prime* as we have learned. Priming is dependent upon connections between ideas/concepts and is for the most part an automatic process. The stimulus that follows the prime is called the *target*. Sometimes this influence is beneficial, as when a prime makes the target easier or faster to process. This kind of positive influence on processing is called facilitation. Occasionally, the influence is negative, as when a prime is misleading. When the prime slows down performance to the target, the negative influence on processing is called inhibition, backward masking, or suppression. Priming is an automatic cognitive process and fundamental to retrieving information from semantic memory, thus the key to facilitation versus inhibition is semantic relatedness, as we have learned. For instance, Loftus and Loftus (1975) found that human performance was significantly better when a known concept was used as a prime (e.g., fruit—apple) than when that letter or adjective was used as the prime (D—mammal; red—fruit). The category name activates its semantic representation and then primes other members of the category.

When letters or adjective targets are presented in an experiment, priming from the category name makes it easier to access a memory of the category, because the memory had already been primed. Conversely, receiving a letter or adjective as a prime has little effect. No relevant activation of potential targets is available with such primes, so there is no facilitation of the word naming. It is not necessary for people to access the meanings of words in a lexical decision task; instead, they need only look up the words in a "mental dictionary" or lexicon, determining if the word has been retrieved from long-term memory—this is a fully automatic process. The network of interpretations reveals the relationship between our semantic concepts and the words we use to name them, that is, between the semantic and the lexical entries in memory. There are some complications in this process, however. Lexical feature complexity can affect the speed of processing (latency) in priming concepts, and lexical ambiguity also leads to retrieval difficulties—they can interfere or inhibit. Spreading activation of semantically related information connects all the meanings of an ambiguous word.

The potency of spreading activation depends on at least two factors: the dominance of a particular meaning for a given concept that has different meanings (e.g., wave), and the degree of surrounding context ("wave at the crowd"). With little or no context, meanings are activated to the level that depends on their associative dominance. With richer context, the meaning receives additional automaticity effects from activations, which has a strong influence on information retrieval speed from semantic memory.

Content and Technical Accuracy

On the one hand, a negative effect of our integrative memory tendencies is that the separate bits of information we store away may not exactly match our existing knowledge, leading to certain kinds of distortions when we attempt to remember things. Later memory retrieval may be technically inaccurate because when we re-create memories; they are a product of both integration and summarization, and not a verbatim recollection. When in a more thoughtful and controlled process, a person may actively suppress and reject information that they retrieve, believing it to be irrelevant. This can present a problem when people must re-create an experience (an episode) exactly, such as demanded in court cases using eyewitness testimonies.

On the other hand, an overwhelming positive aspect of this tendency is that it enhances content accuracy, that is, we can understand and remember the essence of complex meaningful events and episodes. Tasks involving episodic memory generally rely only on performance with the items presented as stimuli. This is a very data-driven process. Subjects involved in episodic task research, in which nothing like the connected meaningfulness of a paragraph is presented to them, recall what they are shown. We call this technical accuracy—subjects must recall or recognize to some degree of accuracy, where accuracy is defined as recalling or reconstructing exactly what was experienced. In more semantic tasks, we find very little emphasis on what people remember exactly about what was presented, but instead focus on remembering the material in terms of its overall meaning.

2.5 Summary

We have now given a broad view of concepts of previous and current research into cognition, and how we have to think of harnessing this knowledge to create systems that work better with human use in the future. When we think of the human and how innovation should center on how the human mind and body works, better products with better efficiencies will be gained. The journeys ahead to alleviate human–computer interaction hurdles, such as cognitive overload, lies in an intimate understanding of who we are, what make us "tick," and what are the best ways to compliment the user and not hinder their natural processing abilities. The human and the machine should act in as much a symbiotic state as possible. Who is to say where we will be in 20, 30, or even 200 years from now in regard to how we interact with technologies. The potential is only as far as we can dream, as we have seen with past conceptualizations in science fiction and other dreams of storytellers that have become a reality in our everyday lives.

The focus of the following chapters is on semantic cognition and what it means for development of computing technology. There will be much discussion and debate, but one thing is clear: We can move forward, and by a basic understanding of human functioning on a cognitive scale, that progression is made easier.

References

Adams, R. (2007). Decision and stress: Cognition and e-accessibility in the information workplace. *Universal Access in the Information Society, 5*(4), 363–379. doi:10.1007/s10209-006-0061-9.

Agentis Software. (2008). *Goal directed agent technology: A whitepaper*. Atlanta GA.

Anderson, J. R. (1983). *The architecture of cognition*. Cambridge: Harvard University Press.

Anderson, J. R. (2000). *Learning and memory: An integrated approach* (2nd ed., Vol. xviii). Hoboken: Wiley.

Averbach, E., & Sperling, G. (1961). Short term storage of information in vision. In C. Cherry (Ed.), *Information theory* (pp. 196–211). Washington, DC: Butterworth & Co.

Baddeley, A. D., & Hitch, G. J. (1974). Working memory. In G. Bower (Ed.), *The psychology of learning and motivation: Advances in research and theory* (pp. 47–90). New York: Academic.

Bandura, A., Adams, N. E., & Beyer, J. (1977). Cognitive processes mediating behavioral change. *Journal of Personality and Social Psychology, 35*(3), 125–139. doi:10.1037/0022-3514.35.3.125.

Bargh, J. A., & Morsella, E. (2008). The unconscious mind. *Perspectives on Psychological Science, 3*(1), 73–79. doi:10.1111/j.1745-6916.2008.00064.x.

Barnhardt, T. (2005). Number of solutions effects in stem decision: Support for the distinction between identification and production processes in priming. *Memory, 13*(7), 725–748. doi:10.1080/09658210444000368.

BBN Technologies. (2004). Cougaar architecture. http://www.cougaar.org. Accessed 18 April 2015.

Bertalanffy, L. V. (1968) *General system theory: Foundations, development, applications*. New York: George Braziller.

Becker, M. W., Pashler, H., & Anstis, S. M. (2000). The role of iconic memory in change-detection tasks. *Perception, 29*(3), 273–286.

Blume, B. D., Baldwin, T. T., & Ryan, K. C. (2013). Communication apprehension: A barrier to students' leadership, adaptability, and multicultural appreciation. *Academy of Management Learning & Education, 12*(2), 158–172. doi:10.5465/amle.2011.0127.

Bransford, J. D., & Johnson, M. K. (1972). Contextual prerequisites for understanding: Some investigations of comprehension and recall. *Journal of Verbal Learning and Verbal Behavior, 11*(6), 717–726. doi:10.1016/S0022-5371(72)80006-9.

Brünken, R., Steinbacher, S., Plass, J. L., & Leutner, D. (2002). Assessment of cognitive load in multimedia learning using dual-task methodology. *Experimental Psychology, 49*(2), 109–119. doi:10.1027//1618-3169.49.2.109.

Brünken, R., Plass, J. L., & Leutner, D. (2004). Assessment of cognitive load in multimedia learning with dual-task methodology: Auditory load and modality effects. *Instructional Science, 32*(1–2), 115–132. doi:10.1023/B:TRUC.0000021812.96911.c5.

Card, S. K., Mackinlay, J. D., & Shneiderman, B. (1999). *Readings in information visualization: Using vision to think*. New York: Morgan Kaufmann.

Carroll, J. B. (1993). *Human cognitive abilities: A survey of factor-analytic studies*. Cambridge: Cambridge University Press.

Chomsky, N. (1979). Human language and other semiotic systems. *Semiotica,* 25, 31–44.

Chomsky, N. (1996). *Studies on semantics in generative grammar*. The Hague: Mouton Publishers.

Craik, F. I. M., & Lockhart, R. S. (1972). Levels of processing: A framework for memory research. *Journal of Verbal Learning and Verbal Behavior, 11*(6), 671–684. doi:10.1016/S0022-5371(72)80001-X.

Cooke, N. J., Gorman, J. C., & Rowe, L. J. (2004). *An ecological perspective on team cognition*. Mesa: Cognitive Engineering Research Institute.

Cowan, N. (2000). The magical number 4 in short-term memory: A reconsideration of mental storage capacity. *Behavioral and Brain Sciences, 24,* 87–185.

Endsley, M. R., Bolte, B., & Jones. D. G. (2003). *Designing for situation awareness: An approach to user centered design*. New York: Taylor & Francis.

Engonopoulos, N., Demberg, V., & Sayeed, A. B. (2013). Language and cognitive load in a dual task environment. In Proceedings of the 35th Annual Meeting of the Cognitive Science Society (CogSci August, 2013) (pp. 2249–2254), Berlin, Germany.

Gavrilova, T. A., & Voinov, A. V. (2007). The cognitive approach to the creation of ontology. *Automatic Documentation and Mathematical Linguistics, 41* (2), 59–64.

Gawronski, B., & Bodenhausen, G. V. (2006). Associative and propositional processes in evaluation: An integrative review of implicit and explicit attitude change. *Psychological Bulletin, 132*(5), 692–731. doi:10.1037/0033-2909.132.5.692.

Giddens, A. (1984). The constitution of society: *Outline of the theory of structuration*. Cambridge: Polity Press.

Gioia, D. A., & Poole, P. P. (1984). Scripts in organizational behavior. *Academy of Management Review, 9*(3), 449–459. doi:10.5465/AMR.1984.4279675.

Goldstein, I. L., & Ford, J. K. (2002). *Training in organizations*. Belmont: Wadsworth.

Greene, E. (2007). Information persistence in the integration of partial cues for object recognition. *Perception & Psychophysics, 69*(5), 772–784. doi:10.3758/BF03193778.

Haber, R. N. (1983). The impending demise of the icon: A critique of the concept of iconic storage in visual information processing. *The Behavioral and Brain Sciences, 6,* 1–54.

Halford, G. S., Baker, R., McCredden, J. E., & Bain, J. D. (2005). How many variables can humans process? *Psychological Science, 16*(1), 70–76. doi:10.1111/j.0956-7976.2005.00782.x.

Hazeltine, E., Ruthruff, E., & Remington, R. W. (2006). The role of input and output modality pairings in dual-task performance: Evidence for content-dependent central interference. *Cognitive Psychology, 52*(4), 291–345. doi:10.1016/j.cogpsych.2005.11.001.

Hull, C. L. (1932). The goal-gradient hypothesis and maze learning. *Psychological Review, 39*(1), 25–43. doi:10.1037/h0072640.

Hutchins, S. D., Wickens, C. D., Carolan, T. F., & Cumming, J. M. (2013). The influence of cognitive load on transfer with error prevention training methods a meta-analysis. *Human Factors: The Journal of the Human Factors and Ergonomics Society,* 0018720812469985. doi:10.1177/0018720812469985.

Irwin, D., & Thomas, L. (2008). Neural basis of sensory memory. In S. Luck & A. Hollingworth (Eds.) *Visual memory* (pp. 32–35). New York: Oxford University Press.

Jacoby, L. L. (1991). A process dissociation framework: Separating automatic from intentional uses of memory. *Journal of Memory and Language, 30*(5), 513–541. doi:10.1016/0749-596X(91)90025-F.

Jensen, P. S., Mrazek, D., Knapp, P. K., Steinberg, L., Pfeffer, C., Schowalter, J., & Shaprio, T. (1997). Evolution and revolution in child psychiatry: ADHD as a disorder of adaptation. *Journal of the American Academy of Child & Adolescent Psychiatry, 36*(12), 1672–1681. doi:10.1097/00004583-199712000-00015.

Killmer, K. A., & Koppel, N. B. (2002). So much information, so little time: Evaluating web resources with search engines. *Technological Horizons in Education Journal, 30,* 21–29.

Khemlani, S., & Johnson-Laird, P. N. (2013). Cognitive changes from explanations. *Journal of Cognitive Psychology, 25*(2), 139–146. doi:10.1080/20445911.2012.720968.

Kozma, R. B. (1991). Learning with media. *Review of Educational Research, 61*(2), 179–212.

Langer, S. (1957). *Philosophy in a new key: A study in the symbolism of reason, rite, and art.* Cambridge: Harvard University Press.

Lassila, O., & Swick, R. R. (1999). Resource description framework (RDF) model and syntax specification. W3C REC-rdf-syntax-19990222. http://www.w3.org/TR/1999/REC-rdf-syntax-19990222/. Accessed 18 April 2015.

Lidwell, W., Holden, K., & Butler, J. (2003). *Universal principles of design.* Rockport Publishers. http://www.amazon.ca/exec/obidos/redirect?tag=citeulike09–20&path=ASIN/1592530079. Accessed 18 April 2015.

Loftus, G. R., & Loftus, E. F. (1975). The influence of one memory retrieval on a subsequent memory retrieval. *Memory and Cognition, 2*(3) 467–471.

Logan, G. D., & Cowan, W. B. (1984). On the ability to inhibit thought and action: A theory of an act of control. *Psychological Review, 91*(3), 295–327.

Lord, R. G., & Maher, K. J. (1990). Alternative information-processing models and their implications for theory, research, and practice. *Academy of Management Review, 15*(1), 9–28. doi:10.5465/AMR.1990.4308219.

Lord, R. G., & Maher, K. J. (2002). *Leadership and information processing: Linking perceptions and performance.* London: Routledge.

McBride, B. (2004). The resource description framework (RDF) and its vocabulary description language RDFS. In P. D. S. Staab & P. D. R. Studer (Eds.), *Handbook on ontologies* (pp. 51–65). Berlin: Springer. http://link.springer.com/chapter/10.1007/978-3-540-24750-0_3. Accessed 18 April 2015.

McNally, R. J., Otto, M. W., Hornig, C. D., & Deckersbach, T. (2001). Cognitive bias in panic disorder: A process dissociation approach to automaticity. *Cognitive Therapy and Research, 25*(3), 335–347. doi:10.1023/A:1010740617762.

Morris, J. S., Ohman, A., & Dolan, R. A. (1999). A subcortical pathway to the right amygdala mediating unseen fear. *Proceedings of the National Academy of Sciences of the United States of America,* 96, 1680–1685.

Paas, F., Renkl, A., & Sweller, J. (2003). Cognitive load theory and instructional design: Recent developments. *Educational Psychologist, 38*(1), 1–4. doi:10.1207/S15326985EP3801_1.

Reder, L. M., & Schunn, C. D. (1996). Metacognition does not imply awareness: Strategy choice is governed by implicit learning and memory. In L. M. Reder (Ed.), *Implicit Memory and Metacognition.* Hillsdale: Lawrence Erlbaum.

Richardson-Klavehn, A., Gardiner, J. M., & Ramponi, C. (2002). Level of processing and the process-dissociation procedure: Elusiveness of null effects on estimates of automatic retrieval. *Memory, 10*(5–6), 349–364. doi:10.1080/09658210244000180.

Rosch, E. (1975). Cognitive reference points. *Cognitive Psychology, 7*, 532–547.

Schacter, D. L. (1995). Implicit memory: A new frontier for cognitive neuroscience. In *The cognitive neurosciences* (pp. 815–824). Cambridge: The MIT Press.

Schacter, D. L., Gilbert, D. T., & Wegner, D. M. (2011). *"Semantic and episodic memory". Psychology* (2nd ed., pp. 240–241.23). New York: Worth, Incorporated.

Seitz, A. R. (2013). Cognitive neuroscience: Targeting neuroplasticity with neural decoding and biofeedback. *Current Biology, 23*(5), R210–R212. doi:10.1016/j.cub.2013.01.015.

Shabata H., & Omura, K. (2012). Comparing paper books and electronic media in reading to answer questions. *Information Processing, Society for Imaging Science and Technology, 1*, 43–46.

Shibata, K., Chang, L.-H., Kim, D., Náñez, J. E., Sr, Kamitani, Y., Watanabe, T., & Sasaki, Y. (2012). Decoding reveals plasticity in V3 A as a result of motion perceptual learning. *PLoS ONE, 7*(8), e44003. doi:10.1371/journal.pone.0044003.

Shibata, K., Watanabe, T., Sasaki, Y., & Kawato, M. (2011). Perceptual learning incepted by decoded fMRI neurofeedback without stimulus presentation. *Science, 334*(6061), 1413–1415. doi:10.1126/science.1212003.

Shiffrin, R. M., & Schneider, W. (1977). Controlled and automatic human information processing: II. Perceptual learning, automatic attending and a general theory. *Psychological Review, 84*(2), 127–190. doi:10.1037/0033-295X.84.2.127.

Shirom, A. (2003). Job-related burnout: A review. In J. C. Quick & L. E. Tetrick (Eds.), *Handbook of occupational health psychology* (pp. 245–264). Washington, DC: American Psychological Association.

Stein, B. E., Wallace, M. T., & Alex, M. (1995). Neural mechanisms mediating attention and orientation to multisensory cues. In M. S. Gazzaniga (Ed.), *The cognitive neurosciences* (pp. 683–702). Cambridge: MIT Press.

Sternberg, R. J. (1977). *Intelligence, information processing, and analogical reasoning: The componential analysis of human abilities* (Vol. xi). Oxford: Lawrence Erlbaum.

Sweller, J. (1988). Cognitive load during problem solving: Effects on learning. *Cognitive Science, 12*(2), 257–285. doi:10.1207/s15516709cog1202_4.

Thomas, L. E., & Irwin, D. E. (2006). Voluntary eyeblinks disrupt iconic memory. *Perception & Psychophysics, 68*(3), 475–488. doi:10.3758/BF03193691.

Trafton, J. G., & Trickett, S. B. (2001). Note-taking for self-explanation and problem solving. *Human-Computer Interaction, 16*(1), 1–38. doi:10.1207/S15327051HCI1601_1.

Tulving, E. (1972). Episodic and semantic memory. In E. Tulving & W. Donaldson (Eds.), *Organization of memory* (pp. xiii, 311–403, 423). Oxford: Academic.

Urakawa, T., Inui, K., Yamashiro, K., Tanaka, E., & Kakigi, R. (2010). Cortical dynamics of visual change detection based on sensory memory. *NeuroImage, 52*(1), 302–308. doi:10.1016/j.neuroimage.2010.03.071.

Watson, D., & Tharp, R. (2013). *Self-directed behavior: Self-modification for personal adjustment*. Cengage Learning.

Zaccaro, S. J., Rittman, A. L., & Marks, M. A. (2001). Team leadership. *The Leadership Quarterly, 12*(4), 451–483. doi:10.1016/S1048-9843(01)00093-5

Chapter 3
Using Symbols for Semantic Representations: A Pilot Study of Clinician Opinions of a Web 3.0 Medical Application

Michael Workman

3.1 Introduction

Technologies such as Resource Description Framework (RDF), domain ontologies, NoSQL, sophisticated search techniques, reasoners, and analytics have greatly improved solutions to "big data" problems. However, research into information visualization has lagged far behind these other technologies. Although tremendous importance has been placed on visual displays with regard to physical layout and the encapsulation of what might be termed "world semantics" (Shneiderman 1992), they have neglected relational and contextual aspects that facilitate meaning making or what might be called display semantics (Bederson et al. 2002). This is particularly true of high-density displays, such as those often found in health care.

Physicians must evaluate the complex relationships among indicators of illnesses, symptoms, laboratory information, and results of cases to diagnose acute patient conditions and decide on treatments. The consequences from misinterpretations or information overload from poorly rendered display media can be devastating as was noted when physicians removed the wrong kidney from a patient in a surgical operation (Dyer 2002).

Many, if not most, clinical displays of medical information in use today render linear forms of media, such as texts, line graphs, and charts, which are inefficient (Lohr 2003). Some previous research has explored replacing conventional linear renderings with more holistic information such as glyphs and graphical linguistics in the medical arena, but these have had mixed results (Workman 2008; Yost and North 2005). Thus, to help advance our understanding of how and why some display media work better than others in a medical setting, we conducted an exploratory study of an in-use medical display technology with a comparable symbol-based

I would like to thank Dr. Michael F. Lesser, MD, for his assistance in the development of the semantic medical displays.

M. Workman (✉)
Advanced Research and Development, Security Policy Institute, Melbourne, FL, USA
e-mail: workmanfit@yahoo.com

M. Workman (ed.), *Semantic Web*, DOI 10.1007/978-3-319-16658-2_3

technology at a hospital in central Florida (USA). Theory-grounded empirical research into medical displays should help move the current largely subjective and anecdotal body of literature toward a greater understanding of what constitutes more effective display designs (Loft et al. 2007; Bradshaw 2006).

Our study used a point-of-care application that displayed all relevant information at the patient's bedside on a fixed display, and on caregiver's mobile devices. The application included all of the functionality to automate the patient care process using the terminology the hospital had validated. Performance thresholds were set to drive care priorities based on the displays. The intent of the designs was to enable the caregiver to glance at all of the patients' vital information simultaneously to gain situational awareness, and determine what the most important next activity should be. In the sections that follow, we briefly present our theoretical foundation along with design principles, then we describe our study in more detail, and finally we present our results and conclusions.

3.2 Theory Foundations and Design Principles

Visual perception occurs on several levels depending on one's focus. For example, normal vision encompasses approximately 60° (with peripheral vision extending out about 20° on either side of the eye), but narrows to between 6 and 10° when focused on an object. Furthermore, since only a fraction of the original light energy from an environment is registered on the retina and the rest is absorbed and diffused by the fluid and structures within the eye, once focused, the optic nerves are more sensitized to moving and changing (e.g., color) objects over stationary and static ones, which helps the perceptual processing in the visual cortex to distinguish the object from foreground, background and parallel objects, and the "meaning" it conveys or that is interpreted (e.g., predator or prey; Doneva and De Fockert 2014).

Known as feature detection, visual stimuli such as lines, edges, angles, and movements are differentially perceived. A feature is a pattern or fragment or component that can appear in combination with other features across a wide range of stimulus patterns. Unelaborated features (those without surrounding context) are difficult to discern. This becomes clear when we consider that we read printed text by first extracting individual features from the patterns, then combining the features into recognizable letters, and finally combining the letters to identify the words. Moreover, words without additional context are often "meaningless." With surrounding context, we use the cognitive heuristic of "likeness" to infer correct from misspelled words in our language if there is enough context from which to make an inference (Elliott et al. 2014).

As an example, most English speaking people are unable to see that a single word in isolation such as "slevin" is misspelled unless we write (in the USA) that "four score and slevin years ago" (a segment from the Gettysburg Address, Abraham Lincoln). Noteworthy is that even with this added context, nationalities other than in the USA may not understand. We might further highlight that "wave" may not make sense unless with the context that we should "wave to the crowd" versus "let's catch the next wave"—because this latter idiom is something easily misunderstood even in the USA unless living near an ocean where "surfing the wave" is a common cognitive script or schema to prime "wave" in experiential context. Therefore, influences of surrounding information along with a person's own previous knowledge are critically important to understand visual information (Khemlania and Johnson-Laird 2013).

Principles 1a–c: (a) A medium must present a visual stimulus in a small area and (b) under urgent conditions "change" to focus one's attention on the area, and (c) must have sufficient situated context for an objective interpretation.

When we read linear information such as prose, we get the sense that our eyes consume the visual information in a continuous fashion. However, the eye sweeps from one point to another in movements (saccades), then pauses or fixates while the eye encodes the visual information. Although this eye movement is fairly rapid (about 100 ms), it takes about twice that long to trigger the movement itself. Next, during the saccade, there is suppression of the normal visual processes and, for the most part, the eye only takes in visual information during the fixation period (roughly 20 ms; Elliot et al. 2014).

This means that there is enough time for about three complete visual cycles of fixation-then-saccade per second. Each cycle of the process registers a distinct and separate visual image, although, generally speaking, the scenes are fairly similar and only a radical shift in gaze would make one input completely different from the previous one. Another important characteristic is that visual information is only briefly stored in iconic memory. The duration of time that an image persists in memory beyond its physical duration depends on the complexity of the information that is absorbed during the encoding process (Doneva and De Fockert 2014).

Principles 2a–b: (a) Encoding and comprehension from linear information is cognitively uneconomical and inappropriate for time-sensitive decision-making from complex data, and thus (b) when practical, complex or dense data should be presented in holistic forms.

Studies (e.g., Endsley et al. 2003) show that there are differences in visual perception between viewing a natural "world" environment versus a computer screen. For one thing, the focus of our field of vision is narrower when working with a computer screen than attending to visual stimuli in a natural environment. Aside from that, part of what makes this feature interesting comes from the notion of encoding specificity, where people store not only the information they are taught but also the environment in which the information was learned.

For instance, studies (e.g., Tulving and Thomson 1973) have shown that students perform better on tests when they take the test in the same classroom where they learned the information compared to when they take the test in a different classroom. Going further, relative to computer displays versus a natural environment, no matter how information may appear on a computer screen (even an image rendered in high-definition 3D), the screen can only display on a flat surface (at present). In sum, people tend to perceive natural stimuli more quickly and effectively than they do in artificial settings. Together, these characteristics are referred to as environmental (or ecological) dimensionality, which suggests that display media should most reflect a natural ecology (Doneva and De Fockert 2014).

Principle 3: Displays should take into account (and incorporate from) the ecosystem in which the information is normally or frequently situated.

Averbach and Sperling (1961) performed a series of interesting experiments that showed, on average, people have deterioration in visual recollection as the information complexity increases. For example, when up to four items were presented in their studies (a "chunk"), subjects' recollection was nearly complete, but when up to 12 items were presented, recollection deteriorated to only a 37% level of accuracy. Furthermore, they found that this poor level of accuracy remained essentially the same even for exposures of the visual stimuli lasting for a long time—in a visual sense (about 500 ms). Consequently, in general, people have a span of visual apprehension consisting of approximately five items presented simultaneously. Newer studies (cf. Cowan 2000; Halford et al. 2005) have found some variability relative to visual persistence based on complexity, and the contrast and background upon which images were rendered.

When dark fields were presented before and after a visual stimulus, visual memory was enhanced (just as a lightning bolt was more visible in a nighttime storm than a daytime storm because of the contrast illumination). Furthermore, these studies indicated that over 50% of items presented were recalled well after a 2-s delay when dark fields were used but, in contrast, accuracy dropped to 50% after only a quarter of a second when light fields were used. Finally, because of backward masking between stimuli, an "erasure" stimulus should be presented in order for the visuospatial sketch pad (VSSP) to rid iconic memory of the previously rendered image because a later visual stimulus can drastically affect the perception of an earlier one (Barnhardt 2005). Although in some cases this can be helpful, such as it can facilitate the priming of information that might appear next (proactive interference), it often creates false perceptions and illusions.

Principles 4a–c: (a) Displays should limit the number of rendered concepts to a "chunk" at a time, (b) should use a light image on a dark background, and (c) before a new image rendering, display an erasure stimulus.

3.3 Method

Participants and Preparations

We enlisted clinical caregivers from a regional hospital in central Florida (USA) to participate in our study. The hospital was being acquired by a large hospital chain and was tasked with evaluating new clinical systems. There were 42 clinicians who took part in the study.

We developed an active symbol display from their currently used ("regular") medical informatics system for patient clinical information. The regular system, provided by one of the top five commercial vendors in the USA, presented mainly linear data displays of information such as line graphs, charts, and text messages. Messages, such as patient status would change color (e.g., from black text to red text) to indicate a condition such as an abnormal laboratory result. We created 25 cases for rendering patient information in both the regular system and a new "symbolic" system. Participants were then given a short (1 h) training session on the symbolic system.

Instrumentation

To address the first, second, and forth set of principles, the symbolic system utilized familiar glyphs along with color changes, augmentation such as illuminating a circle around a glyph when there was a change in status, and highlighting (see Fig. 3.1).

For example, a green icon for an empty surgical table indicated that a surgery was scheduled and a surgical room was ready. A yellow icon indicated in transit, for example, a yellow surgical table along with a directional arrow indicated that a patient was either in transit from surgery or to surgery depending on the direction of the arrow. A white surgical table with a patient on the table indicated that the patient was in surgery. Other symbols included pharmacy orders, laboratory tests, X-rays, and so forth. These had modifiers, for example, when a laboratory test was ordered, a microscope was displayed. When the test was completed, a circle was illuminated around the microscope. If the laboratory test was abnormal, it would change color to red, and optionally vibrate. Other indicators included patient at risk for a fall,

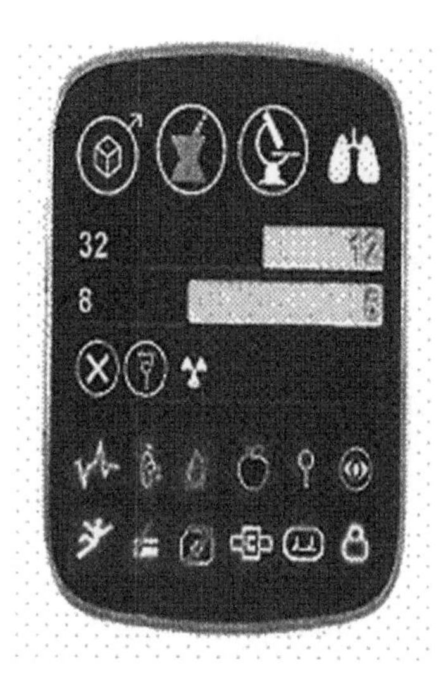

Fig. 3.1 Symbolic display

aspiration risk, nonambulatory, etc. Finally, there were length-of-stay indicators for standard of care.

To address principle 3, the symbols were placed on a "mobile phone" metaphor such that the display appeared exactly the same whether on a mobile phone, a tablet, or large computer screen. The only difference among these was the number of phone metaphor displays that could be shown on a device. For example, 60 displays were possible on a 46-in. monitor. With the symbolic system, at a glance, a clinician could ascertain situational awareness, for instance, that a patient was male, had pharmacy and laboratory work ordered, was at risk for a fall, was a third of the way through a treatment, and vitals appeared normal. For equivalent situational awareness using the regular (linear) system, a large amount of screen real estate was required, cognitive processing of a large number of data points and linear line graph data were factored.

Procedures

For the study, a computer program was written to present cases in random order, but alternating between a regular display and a symbolic display. The program rendered each case for 10 s, after which a list of five descriptions was presented for the participant to select the case event (e.g., male patient at risk for fall has abnormal laboratory result returned, white blood cells (WBC) elevated). At the end of the session, participants answered an opinion questionnaire on a 7-point Likert scale comparing the regular and symbolic displays based on time to interpret, goal facilitation, effectiveness, quality, and potential to prevent interpretation errors.

3.4 Results

We tested the opinion data using nonparametric statistics and the accuracy results using dependent *t* test. The descriptive statistics for the display are shown in Table 3.1.

Compared to the regular display, participants felt that the symbolic system saved them time ($\chi^2=16.57$, $p<0.01$), better facilitated their goals ($\chi^2=17.43$, $p<0.01$),

Table 3.1 Descriptive statistics

	N	Mean	Standard deviation
Time	42	5.59	1.02
Goals	42	4.53	1.22
Effective	42	4.56	1.47
Quality	42	4.81	0.94
Errors	42	4.33	0.99

was more effective in rendering information ($\chi^2=28.33$, $p<0.001$), was of higher quality ($\chi^2=37.76$, $p<0.001$), and had a great potential to reduce errors ($\chi^2=44.85$, $p<0.001$). The timed test for accuracy indicated that participants were better able to accurately determine the actual case condition ($t=-4.79$, $p<0.001$, $\mu=-3.69$, $\sigma=5.00$). The questionnaire allowed for comments, which we present in the "Discussion" section that follows.

3.5 Discussion

Although our pilot study was far from explanatory, it does lend empirical support for the theoretical foundation. Since much of the human factors literature on information displays has been qualitative and anecdotal, more theory-grounded research, especially drawing from cognitive and visual physiology, is needed to inform practice. This has been particularly the case in medical informatics, where the sophistication of display media has lagged far behind other technologies such as medical ontologies, semantic integration, and reasoning and inferential systems.

Participants in our study indicated that the display features in the symbolic system may save up to 2 h per shift per clinician by reducing time required to examine the textual and linear data and search multiple displays for relevant information. They felt also that the use of shapes and color was more effective in highlighting the conditions over the stationary and static line graphs and pie charts utilized by the regular system. In addition, they felt that they were able to distinguish the features much more readily than the simple color changes in the text in the regular system. Beyond these attributes, respondents commented that the symbolic display would enable better communications across disparate groups resulting in process improvements; an example given was that particular metrics such as discharge time would improve because team members from disparate groups who would influence the results would be able to view the same data.

Symbolic systems are not new; however, what has only been fairly recently appreciated is that symbols can convey semantics when augmented with contexts and even attributes that resemble vocabularies. One advantage of symbolic systems is that more information can be conveyed in a reduced area compared to linear and textual information. Another advantage is that they may help reduce cognitive information overload because they present information more holistically and are more cognitively economical. Continued research is needed into developing ways of rendering symbols that can be objectively interpreted, since in many if not most instances the glyphs lack a well-defined, universal, or standard grammar—i.e., they cannot be juxtaposed in relational or subject–predicate forms that can be related across broad areas. Combining the display layer with underlying medical lower domain ontology may help resolve the issue, but this remains unfinished business with many opportunities for the future in semantic visualization.

References

Averbach, E., & Sperling, G. (1961). Short-term storage of information in vision. In C. Cherry (Ed.), *Information theory* (pp. 196–211). London: Butterworth.

Barnhardt, T. M. (2005). Number of solutions effects in stem decision: Support for the distinction between identification and production processes in priming. *Memory, 13,* 725–748.

Bederson, B. B., Shneiderman, B., & Wattenberg, M. (2002). Ordered and quantum treemaps: Making effective use of 2D space to display hierarchies. In B. B. Bedderson & B. Shneiderman (Eds.), *The craft of information visualization* (pp. 257–278). San Francisco: Morgan Kaufmann.

Bradshaw, L. (2006). *Information overload and the Hurricane Katrina post-disaster disaster.* Fremantle: Information Enterprises.

Cowan, N. (2000). The magical number 4 in short-term memory: A reconsideration of mental storage capacity. *Behavioral and Brain Sciences, 24,* 87–185.

Doneva, S. P., & De Fockert, J. W. (2014). More conservative go/no-go response criterion under high working memory load. *Journal of Cognitive Psychology, 26*(1), 110–117.

Dyer, C. (2002). Doctors go on trial for manslaughter after removing wrong kidney. *British Medical Journal, 324,* 10–11.

Elliott, E. M., Morey, C. C., Morey, R. D., Eaves, S. D., Shelton, J. T., & Lutfi-Proctor, D. A. (2014). The role of modality: Auditory and visual distractors in Stroop interference. *Journal of Cognitive Psychology, 26*(1), 15–26.

Endsley, M. R., Bolte, B., & Jones, D. G. (2003). *Designing for situation awareness: An approach to user-centered design.* New York: Taylor & Francis.

Halford, G. S., Baker, R., McCredden, J. E., & Bain, J. D. (2005). How many variables can humans process? *Psychological Science, 16,* 70–76.

Khemlania, S., & Johnson-Laird, P. N. (2013). Cognitive changes from explanations. *Journal of Cognitive Psychology, 25*(2), 139–146.

Loft, S., Sanderson, P., Neal, A., & Mooij, M. (2007). Modeling and predicting mental workload in en route air traffic control: Critical review and broader implications. *The Journal of the Human Factors and Ergonomics Society, 49,* 376–399.

Lohr, L. (2003). *Creating visuals for learning and performance: Lessons in visual literacy.* Upper Saddle River: Prentice Hall.

Shneiderman, B. (1992). *Designing the user interface: Strategies for effective human-computer interaction.* Boston: Addison-Wesley Longman.

Tulving, E., & Thomson, D. (1973). Encoding specificity and retrieval processes in episodic memory. *Psychological Review, 80*(5), 352–373.

Workman, M. (2008). An experimental assessment of semantic apprehension of graphical linguistics. *Computers in Human Behavior, 24,* 2578–2596.

Yost, B., & North, C. (2005). *Single complex glyphs versus multiple simple glyphs.* Proceedings from the Conference on Human Factors in Computing Systems, Portland, OR, USA. April 2–7, pp. 1889–1892.

Chapter 4
Emerging Semantic-Based Applications

Carlos Bobed, Roberto Yus, Fernando Bobillo, Sergio Ilarri, Jorge Bernad, Eduardo Mena, Raquel Trillo-Lado and Ángel Luis Garrido

In the last decade, we have witnessed the birth and spread of the so-called Semantic Web. From its initial proposal by Tim Berners-Lee (Berners-Lee et al. 2001; Shadbolt et al. 2006), to the latest trends and initiatives, such as Linked Data (Bizer et al. 2009a) and DBpedia (Bizer et al. 2009b; Mendes et al. 2012), the Semantic Web is progressively changing the landscape of the World Wide Web (WWW) through the use and adoption of the different semantic technologies that have come along with it. We can see how, although some of the goals of the Semantic Web have not been reached yet, several well-known and successful applications are already using semantic technologies, such as Google's Knowledge Graph, Microsoft's Satori, or Facebook's Graph Search.

In this successful scenario, ontologies have played a crucial role. Defined by Tom Gruber as "an explicit specification of a conceptualization" (Gruber 1993, 1995), ontologies allow to model and capture the semantics of different knowledge domains, providing a means to share definitions, and reach an implicit agreement on the meaning of the published information. Ontologies represent the vocabulary of some domain from a common perspective using a formal language, such as the current standard Web Ontology Language (OWL 2; Hitzler et al. 2012).

Being knowledge representation frameworks as they are, they might have an impact on many other different kinds of systems. In fact, they have already had; for example, ontologies have been successfully used in the integration of information and information systems (Mena and Illarramendi 2001; Wache et al. 2001). Thus, with their advance and the development of their associated technologies, we can

C. Bobed (✉) · R. Yus · F. Bobillo · S. Ilarri · J. Bernad · E. Mena · R. Trillo-Lado · Á. L. Garrido
Department of Computer Science & Systems Engineering, University of Zaragoza, C. María de Luna 1, 50018 Zaragoza, Spain

C. Bobed · R. Yus · F. Bobillo · S. Ilarri · J. Bernad · E. Mena · R. Trillo-Lado · Á. L. Garrido
e-mail: cbobed@unizar.es; ryus@unizar.es; fbobillo@unizar.es; silarri@unizar.es; jbernad@unizar.es; emena@unizar.es; raqueltl@unizar.es; garrido@unizar.es

M. Workman (ed.), *Semantic Web*, DOI 10.1007/978-3-319-16658-2_4

now explore deeper in the quest for smarter information systems which exploit the semantics of data.

Possible sources of enhancement for our applications could be the use of ontologies which are already shared and published, the possibility of modeling domains in detail, thanks to the expressivity of the different families of languages (mainly based on Description Logics (DLs; Baader et al. 2003)), or the exploitation of the huge amount of semantic data which have been already generated. In this chapter, we present different semantic-based applications and projects that we have developed in the Distributed Information Systems research group (SID, http://sid.cps.unizar.es) that currently benefit from these semantic technologies, and provide a good example of how the addition of semantics broadens the capabilities of an information system. In particular:

- In our approach to what could be considered as the most classical conception of the Semantic Web, we have applied the use of semantics to the field of keyword-based search. Using disambiguation techniques which exploit the knowledge stored in ontologies (Gracia and Mena 2008; Trillo et al. 2007a), we have developed two different approaches to keyword search over different information systems: *QueryGen* (Bobed 2013) and *Doctopush* (Trillo et al. 2011). The former one is oriented to perform semantic keyword-based search over heterogeneous information systems, proposing a generalized semantic keyword interpretation process, while the latter aims at performing semantic data retrieval over the WWW using the semantics of keywords to cluster relevant sources of information.
- We have also devised a framework to enhance semantically different tasks regarding information extraction (IE), such as automatic text classification, semantic search, and summarization of text sources, among others. This framework, called *GENIE* (GENeric Information Extraction Framework), aims at supplying a set of libraries designed to assist developers in projects of this nature, adopting a semantic approach to all modules, thus taking advantage of the latest advances made in ontological engineering and semantics.
- Regarding the standard formalism used for ontologies, we have studied an extension of the semantics of DLs to embrace Fuzzy Logic. This has led to the implementation of the fuzzy ontology editor *Fuzzy OWL 2* (Bobillo and Straccia 2011), and the fuzzy DL reasoners *DeLorean* (Bobillo et al. 2012a) and *fuzzyDL* (Bobillo and Straccia 2008), combining the expressivity of classical DL languages (which use crisp logic) with the flexibility of fuzzy logics for imprecise knowledge representation.
- Finally, we have studied the relationship between locations and semantics. On the one hand, we have applied the know-how acquired using ontological formalisms to study how we can model locations using different granularities while keeping and exploiting their semantics. On the other hand, we are currently working on enhancing the use of Location-Based Services (LBSs) adding semantics, which is crystallizing in *SHERLOCK* (Yus et al. 2014a), a system that searches and shares up-to-date knowledge from nearby devices to provide users with interesting LBSs.

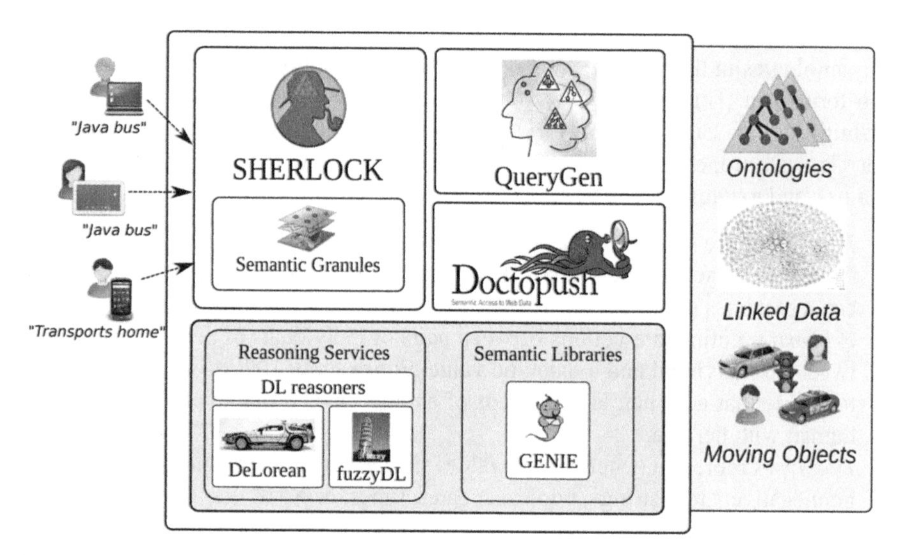

Fig. 4.1 Overview of our semantic-based systems

In fact, all of these systems are not isolated, but help each other to perform different tasks. In Fig. 4.1, we can see how they could be coordinated. SHERLOCK uses our different semantic models of locations to give meaning to the locations of the users, and infer further information out from them. Also, SHERLOCK might provide a keyword interface and could use QueryGen to construct a formal specification of the requested service out from the user's input keywords. Doctopush could be integrated as well, and used as a particular service under the umbrella of SHERLOCK.

All of these systems use several reasoning and semantic services, which could be abstracted in a separated layer. On the one hand, this layer would expose the services provided by classical DL reasoners, as well as the reasoning services of our fuzzy extension for DLs represented in Fig. 4.1. On the other hand, we could find the services exposed by GENIE to enhance different semantic tasks that they have to perform.

For the sake of readability, we devote the next section to give a brief overview of the basics on DL-based ontologies and DL (Baader et al. 2003) reasoners, which are thoroughly used in the presented systems. Then, in the rest of this chapter, we elaborate on these four points, all of which share the use of semantics (at different levels) as their main value.

4.1 Background

Ontologies (Frank 2003; Haarslev et al. 1998) are one of the most popular approaches to represent the knowledge of a domain. Although they can be developed using different languages and formalisms (Gómez-Pérez et al. 2004), we focus on the basics of OWL, the language that is the current W3C standard for representing ontologies in the Semantic Web, which has Description Logics as the underlying formalism. Thus, in DL-based ontologies, the basic ontological representation primitives (also called ontology elements) are individuals (or instances), concepts (or classes), properties (also called roles or relations), datatypes (or concrete domains), and axioms:

- *Individuals* are objects of the world. For instance, *John.*
- *Concepts* are sets of individuals. For example, *Human.* We denote the set of all concept names of an ontology by N_C.
- *Properties* define interactions between pairs of individuals of the domain, or between an individual and a datatype value. For example, *isParentOf* can be used to define that a human is the parent of another one, while *hasAge* can relate a human with her age.
- *Datatypes* represent concrete data values such as numbers (real, rational, integer, nonnegative, etc.), strings, booleans, dates, times, or XML literals, among many other possibilities.
- *Axioms* are formal conditions to be verified by the elements. An ontology can be seen as a finite set of axioms, usually divided in three parts: an assertional box (*ABox*), a terminological box (*TBox*), and a role box (*RBox*), with axioms about individuals, concepts, and roles, respectively. For example, an ABox can assert that *John* is a member of the concept *Human*, and a TBox can assert that *Man* is a subclass of *Human,* usually denoted $Man \sqsubseteq Human$.

Apart from the explicitly represented knowledge, it is possible to perform several reasoning tasks to deduce implicit knowledge, that is, logical consequences of the knowledge in an ontology. This is possible because OWL, as mentioned before, has a formal semantics based on DLs (Baader et al. 2003).

DLs are a family of logics for representing structured knowledge. They are a well-known formalism providing a good trade-off between expressivity of the representation and efficiency of the reasoning. Each DL is denoted by using a string of capital letters which identify its expressivity. For instance, the standard language for ontology representation OWL 2 is equivalent to the DL $\mathcal{SROIQ}(\mathbf{D})$. The expressivity of a DL translates into what kind of constructors can be used to form new concepts. For example, if letter C is in the expressivity of a DL, it means that it can use the constructor $\neg$ to express the contrary of a concept ($Woman \sqsubseteq \neg Man$, a woman cannot be a man). We summarize informally in Table 4.1 the most common constructors that lead to a DL with expressivity $\mathcal{ALC}$, while other logics and their allowed constructors are presented in Table 4.2. For more formal details, see Baader et al. 2003.

The most typical reasoning services in ontologies are designed to check:

- *Consistency:* An ontology O is consistent iff it has a model, that is, there is an interpretation satisfying every axiom in O.
- *Entailment:* O entails an axiom τ iff every model of O satisfies τ.
- *Concept satisfiability:* A concept C is satisfiable w.r.t. O iff it is not interpreted as the empty set in some model of O.

Table 4.1 Constructors and their meanings for $\mathcal{ALC}$ DL

Constructor	Meaning
$\top$	Any element
$\bot$	No element, empty set
A	Atomic concept
¬C	Elements that are not in C
C ⊓ D	Elements in C and D
C ⊔ D	Elements in C or D
∀R.C	Elements "a" such that if "a" is related with "b" by the property R, then "b" is in C
∃R.C	Elements "a" that are related by property R with an element "b" in C

Table 4.2 Expressivity and complexity of reasoning in some important DLs

Logic	Expressivity	Complexity class
$\mathcal{AL}$	$\top, \bot, \neg A, \sqcap, \forall, \exists R.\top$	PTIME
$\mathcal{ALC}(=\mathcal{ALUE})$	$\top, \bot, \sqcap, \sqcup, \forall, \neg, \exists$	EXPTIME
$\mathcal{SHIF}$(**D**) (OWL Lite)	(S =) $\mathcal{ALC}$ + transitive roles, role hierarchies ($\mathcal{H}$), inverse roles ($\mathcal{I}$), functional roles ($\mathcal{F}$), datatypes (**D**)	EXPTIME
$\mathcal{SHOIN}$(**D**) (OWL DL)	$\mathcal{SHI}$, nominals ($\mathcal{O}$), nonqualified numerical restrictions ($\mathcal{N}$), datatypes (**D**)	NEXPTIME
$\mathcal{SROIQ}$(**D**) (OWL 2)	$\mathcal{SHOIQ}$(**D**), complex role inclusion ($\mathcal{R}$), self-restriction, and additional role axioms	N2EXPTIME
$\mathcal{EL}^{++}$ (**D**) (OWL 2 EL)	$\top, \bot, \sqcap, \exists$, role hierarchies, nominals, concrete domains (use of constructors with syntactical restrictions)	PTIME
DL-Lite (OWL 2 QL)	$\top, \bot, \sqcap, \exists, \neg$ (use of constructors with syntactical restrictions)	LOGSPACE
DLP (OWL 2 RL)	$\top, \bot, \sqcap, \sqcup, \forall, \neg, \exists$, cardinality cardinality restriction (0 .. 1) (use of constructors with syntactical restrictions)	PTIME

- *Concept subsumption:* A concept D subsumes a concept C w.r.t. O iff C is interpreted as a subset of D in every model of O.
- *Classification:* The classification of an ontology O consists of computing a hierarchy of concepts based on their subsumption relation.

There are plenty of software applications (called reasoners) implementing ontology reasoning services. Some examples are JFact, HermiT (Glimm et al. 2014), or Pellet (Sirin et al. 2007).

4.2 Semantics Behind Keywords

The usage of keyword search has spread in the past few years thanks to its simplicity and its adoption by the main web search engines. Common users have found in it an easy way to express their information needs, defining their searches just by giving a plain set of keywords, and letting the system do all the work for them. However, the ease of use of keyword search comes from the simplicity of its query model, whose expressivity is low compared with other more complex query models (Kaufmann and Bernstein 2010).

Moreover, the use of keyword queries as starting point for information searches introduces a semantic gap between the user intention and the queries as, in fact, keyword queries are simplifications of the queries that really express the user's information need. Thus, there might be a gap between the posed query and the information that the users would like to obtain, for example, when talking about web searches, users usually have to browse the returned web pages looking for the needed information.

In this context, we also have to bear in mind that polysemous words introduce ambiguity in such queries, which cannot be solved without the intervention of the user. For example, if a user inputs the keyword "apple" in Google, locating information about the fruit with such a name will be difficult for her. (As of June 23, 2014, no hit about the fruit appears in the first 40 ranked positions provided as result, with the notorious exception of the page in Wikipedia, whose results are promoted.) Another different example of ambiguity could be the keywords "apache attack," where a user might be looking for information about the Apache helicopter or about how to secure an Apache server. One could argue that the ambiguity in these examples is due to the lack of input keywords, but experience tells us that the average number of keywords used in keyword-based search engines "is somewhere between 2 and 3" (Manning et al. 2008), which points out another problem of keyword search: Users tend to omit important keywords/information, as they consider them implicit in the query.

Being as useful as keyword-based searches have proved to be, we advocate for improving them by first establishing the proper meaning of each input keyword, which allows knowing exactly what the user is referring to with each of the keywords in the input set. This implies discovering possible meanings for each of the keywords, and disambiguating them to obtain their most probable one separately and as a whole set (i.e., the meaning of a keyword affects to the rest of keywords in the input set). In the rest of this section, we explain how we have applied this approach in two different systems which exploit the semantics behind keywords. First, we present QueryGen (Bobed 2013), a system that performs semantic keyword searches over heterogeneous information systems, interpreting keyword queries to access the underlying systems by taking into account the semantics of both the input keywords and the query languages involved. Then, we present Doctopush (Trillo et al. 2011), a pure Web-based search system, where the semantics of the keywords are used to categorize and group dynamically the results of a keyword query, providing more accurate and relevant information.

Semantic Keyword-Based Search: QueryGen

As we have seen before, using keywords as input language makes a system easier to use, but it implies that the queries that users can pose to the search system are limited by the lack of expressivity of this query model. Keyword queries are simplifications of the queries that really express the user's information need. Moreover, experience tells us that users tend to omit important keywords, as they consider them implicit in the query (recall that the average number of keywords used in keyword-based search engines "is somewhere between 2 and 3" (Manning et al. 2008)). However, the use of expressive formal languages such as SQL (ISO/IEC 2011) or SPARQL (Harris et al. 2013) is far from being easy for common users. What is more, the user must know the underlying schema and data she is accessing to effectively query it.

Thus, the sweet spot would be to mix the expressivity of formal languages with the ease of use of keyword queries, while making the user unaware of the data sources being accessed to solve her information needs. To reach this sweet spot, we advocate for a *semantic keyword-based search*, a keyword-based search process which takes into account the semantics of both keywords and query languages during the whole search process. Our objective is to discover and solve the user's information need taking as starting point a set of input keywords. We divide this task into three sub-objectives:

- To discover the exact meaning of each of the keywords in the set of input keywords.
- To give them an interpretation and express it into a formal language to capture the information need accurately.
- To access the proper information system/s transparently to the user, taking into account the different characteristics that the accessed systems might exhibit.

The first objective, the discovery of each keyword's meaning, allows us to work during the whole process with keywords with well-established semantics, which we call *semantic keywords*. The second one implies structuring a bag of keywords into a structured query, a process which is named *keyword query interpretation* (Fu and Anyanwu 2011). The achievement of the last objective is strongly helped by having the information formally expressed, allowing our system to access semantically even non-semantically enhanced data sources. The semantics behind the input set of keywords, the semantics of the different query languages, and the different semantics of the access models are considered to provide a flexible and efficient way to perform a semantic keyword search on heterogeneous information systems. Figure 4.2 shows the three main steps of the process, presenting further details on the disambiguation step.

Discovery of Keyword Senses First of all, we have to introduce the exact meaning of *sense* in our system: A sense is the precise meaning of a keyword in a context, that is, the meaning of a keyword is determined by its surrounding keywords. In our system, a sense is represented by a tuple formed by the term itself, an ontological context composed by a list of possible synonyms (with their URIs) and ontological

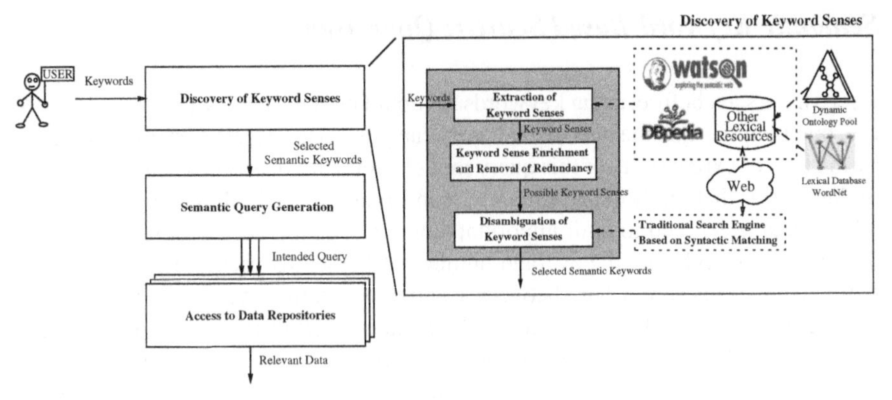

Fig. 4.2 An overview of the whole process: From plain keywords to data access

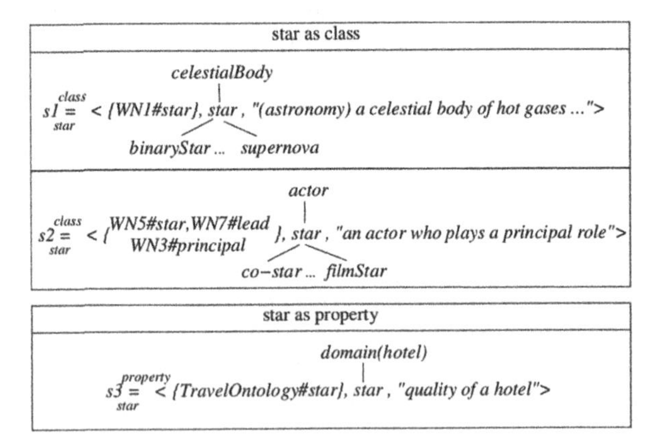

Fig. 4.3 Possible senses for keyword *star*

information about the term, and a description in natural language. Each ontological context is built by integrating information from different ontologies. Figure 4.3 shows some possible senses for the user keyword *star*, retrieved from online ontologies.

So, our search starts by discovering and building these senses for the plain input keywords. Then, the discovery of the semantics behind each of the input keywords is done by taking into account their individual possible semantics as well as the possible semantics of their context (the rest of keywords), following the proposal in Trillo et al. (2007a). In particular, this process is divided into three substeps (see Fig. 4.2):

- *Extraction of Keyword Senses:* The system obtains the possible meanings of each keyword by consulting a dynamic pool of ontologies (in particular, it queries

Watson (d'Aquin et al. 2007), DBpedia (Mendes et al. 2012), WordNet (Miller 1995), and other ontology repositories to find ontological terms that syntactically match the keywords, or one of their synonyms). For each matching, the system builds a sense, which is semantically enriched with the ontological terms of the corresponding synonyms by also searching in the ontology pool. As a result of this step, we obtain a list of candidate keyword senses for each user keyword. In Fig. 4.3, three possible senses (two as a class and one as a property) retrieved for the user keyword *star* are shown.

- *Keyword Sense Enrichment and Removal of Redundancy:* In the sense list obtained in the previous step, there might be redundant meanings as the senses for each keyword are built with terms extracted from different ontologies. An incremental algorithm is used to align the different keyword senses and merge them when they are similar enough, and thus, to avoid redundancy. Our system calculates a *synonymy probability* that considers both linguistic and structural characteristics of the source ontologies: The linguistic similarity is calculated considering the different labels of each term as strings, and the structural similarity is calculated recursively by exploiting the semantics of the ontological context of the keyword sense until a certain depth. Finally, both similarity values are combined to obtain the resultant synonymy measure. The formulae for the synonymy for each type of senses (concepts, roles, and instances) can be found in Trillo et al. (2007a). Senses are merged when the estimated *synonymy probability* between them exceeds a certain threshold.[1] Thus, the result is a set of *different* possible senses for each user keyword entered.
- *Disambiguation of Keyword Senses:* A disambiguation process is carried out to select the most probable intended sense of each user keyword by considering the possible senses of the rest of keywords. The senses are compared by combining (Gracia and Mena 2009): (a) a Web-based relatedness measure that measures the co-occurrence of terms on the Web according to traditional search engines such as Google or Yahoo, (b) the overlap between the words that appear in the context, and the words that appear in the semantic definition of the sense (Banerjee and Pedersen 2003), and (c) the frequency of usage of senses (when available, as in WordNet annotated corpora). Our disambiguation process can be extended by including other different disambiguation algorithms such as the ones defined in Po (2009), as our approach does not depend on a specific disambiguation algorithm. This way, the best sense for each keyword will be selected according to its context. Note that this selection might require the user's feedback to select the most appropriate sense for each keyword in a semiautomatic way.

However, establishing the meaning of each keyword of the input is just the first step to obtain a proper interpretation, as several queries might be represented by a given set of keywords. For example, given the keywords *fish* and *person* meaning "a creature that lives and can breathe in water" and "a human being," respectively,

[1] In Gracia et al. (2009), the authors proposed several strategies to obtain this threshold and validated them via thorough experimentation.

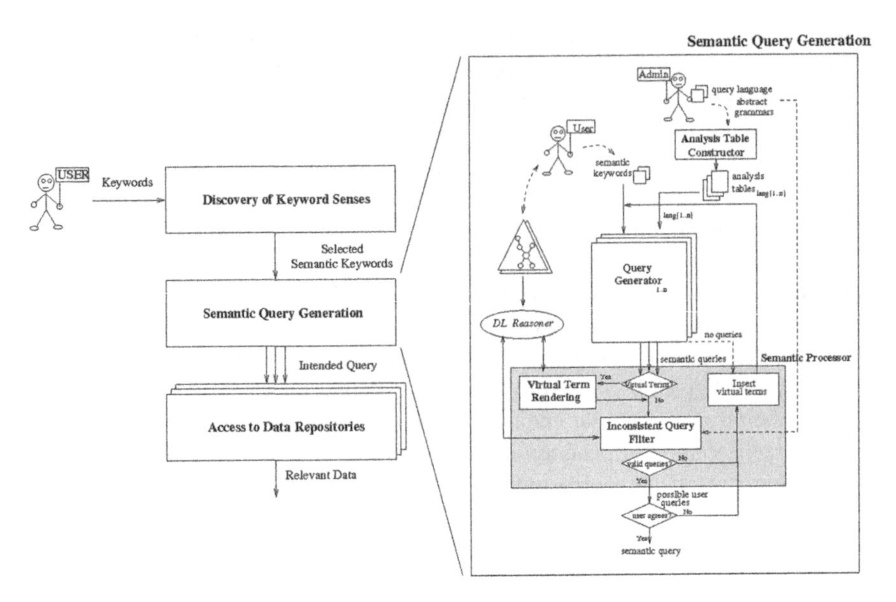

Fig. 4.4 Multi-language query generation process

the user might be asking for information about either biologists, fishermen, or even other possible interpretations based on those individual keyword meanings.

Semantic Query Generation The output of the previous step is a set of keywords which have their meaning properly attached, which we call *semantic keywords*. The ontological information that has been considered for obtaining the meaning of each keyword comes along with each of them. Our system automatically integrates this information, and then automatically builds a set of formal queries which, combining all the keywords, represents the possible semantics that could be intended by the user when she wrote the list of plain keywords. To do so, our system performs the following steps (see Fig. 4.4)

- *Analysis Table Constructor:* In order to capture formally the user's information need, the possible queries are expressed in the different query languages made available to our system, which are modeled using semantically annotated abstract grammars. These grammars lack syntax sugar and define: (1) how to combine the operators of a query language using *typed gaps*, that is, they specify which kind of queries can be built using concepts, roles, and instances in the corresponding query language (e.g., *And concept concept*), and (2) the semantics of the different operators giving their properties (e.g., associativity, symmetry, etc.) and DL expressions that will be checked with the help of a DL reasoner. The construction of the analysis tables (Aho et al. 2006) for the formal query languages is done off-line and just once for each language made available to our system.

- *Query Generator:* With the analysis tables, our system builds the possible queries for each query language according to its syntax. First, the Query Generator builds all the syntactically possible combinations according to the grammars of the available query languages. We call these combinations *abstract queries* because they have *gaps* that will be filled later with specific concepts, roles, or instances. These abstract queries are represented as trees, where the nodes are operators and the leaves are *typed gaps* (concept, role, or instance gaps). Then, for each abstract query tree generated, the gaps in the leaves are filled with the user keywords matching the corresponding gap type (i.e., keywords mapped to concepts are used to fill concept gaps in abstract queries, roles to fill role gaps, and so on). During this generation process, QueryGen takes into account the semantics of the different operators to avoid generating semantically equivalent queries.
- *Semantic Processor:* Then, once the set of syntactically possible queries is obtained, the Semantic Processor filters out the inconsistent ones with the help of a DL reasoner. This is done by using the DL expressions that specify each query language, which enables QueryGen to obtain an expression for the semantic consistency of each of the queries.

When no query satisfies the user, our system also performs a semantic enrichment of the input by adding *virtual terms*. They are generic typed gaps (to be replaced by concepts, roles, or instances) that represent the keywords that the user might have omitted, but without whom the intended query cannot be built. In a new query generation step, our system treats them as regular typed gaps but, instead of being replaced by input keywords, they are substituted by terms obtained from the ontologies which the input keywords were mapped to (during the previous discovery step). Thus, any query that the user could have in mind will be generated as a candidate interpretation as long as the available query languages used are expressive enough. Note how this query generation process has both a syntactic and semantic dimension: It generates only syntactically correct queries according to the grammar of each of the query languages made available to the system, and it takes into account the semantics of the operators of each language and the semantics of the keywords to avoid generating either duplicated or incoherent queries. This process is performed in parallel for each available query language, as their expressivity can differ from each other.

Access to Data Repositories Finally, once the user has validated the generated query that best fits her intended meaning, the system forwards it to the appropriate underlying structured data repositories (databases, Linked Data endpoints, etc.) that will retrieve data according to the semantics of such a query. This is not a trivial task, as our system must be able to adapt itself to their different query processing capabilities and access methods, and to their different data models and formats of the retrieved data. To provide QueryGen with enough flexibility to deal with this data heterogeneity, we advocate for the architecture shown in Fig. 4.5, whose main modules are the following ones:

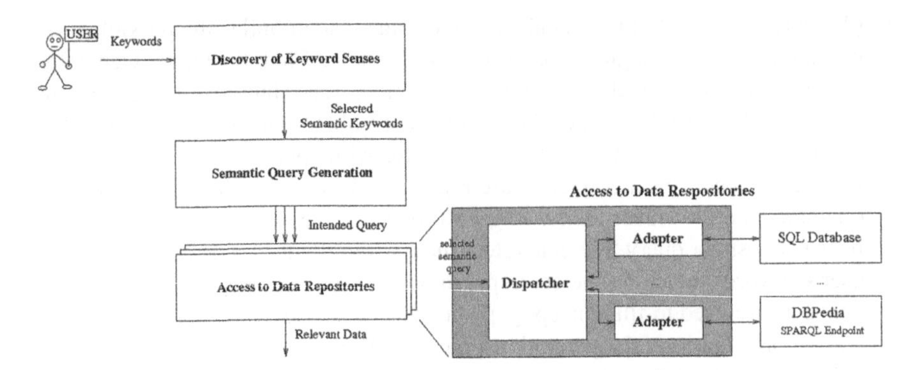

Fig. 4.5 Our system can retrieve data from different channels and data models

- *Dispatcher:* Once the user selects her intended query from those generated by QueryGen, the Dispatcher poses the query to the underlying data repositories that are capable of processing it. Every underlying system that is capable of processing the selected query is accessed in a parallel way as any of them could hold the desired answer. Finally, the Dispatcher correlates the data coming from the different systems and presents them to the user.
- *Adapter:* It wraps the access to the data stored in information systems with a certain data organization (e.g., there is an Adapter for relational databases, a different one for SPARQL endpoints, etc.). It registers itself in QueryGen providing information about the querying capabilities of the accessed information system, and making itself available to the Dispatcher. There is one instance of the appropriate kind of Adapter for each system accessed by QueryGen.

These *Adapters* are an evolution of the notion of *wrappers* used in OBSERVER (Mena and Illarramendi 2001), and encapsulate both the access methods and the actual syntax of the query languages and data formats, allowing QueryGen to abstract from them. Thus, we can add new information systems to feed QueryGen just by implementing and registering an appropriate Adapter in the system.

To sum up, QueryGen adopts an approach to the problem of keyword interpretation which provides a solution that, exploiting the semantics of all the elements that participate all along the search process, is flexible enough to deal with different data schemas (ontologies), different query languages, and different execution semantics. Moreover, using QueryGen, users can turn their information needs into formal queries without having to master the formal languages they are written in. Having formal queries instead of information needs removes the ambiguity and enables the systems to focus on answering the specific query that users would have posed if they knew how to write it. For further details on each of the different aspects, we refer the interested reader to Bobed (2013).

Semantic Data Retrieval: Doctopush

As the WWW evolved and became more and more popular, search services to locate web sites and pages have become indispensable for users. Broadly speaking, these search tools could be categorized (at first) as web directories-based ones, and web search engines. Web directories became less relevant than search engines because they do not scale properly due to the manual process required to classify the web pages and sites. So, the main research efforts were focused on web search engines, especially on those with keyword-based interfaces because of their ease of use and success. However, as mentioned above, the use of keyword queries to start off information searches introduces semantic gaps between the user information needs and the queries.

With the advent of the so-called Web 2.0, this need for efficient search tools increased, as users became content providers who often interact with other Web users, and thus the volume of content of the Web increased exponentially very quickly. To help users, keyword-based search engines specialized in different areas, such as job offers, books, etc., have been created in the last decade. They can be regarded as a hybridization of web directories and web search engines and are called *vertical search engines*. Some popular examples of them are Google Scholar (http://scholar.google.com, to search academic and research articles and papers), Technorati (http://technorati.com, to search blogs), and InfoJobs in Spain (http://www.infojobs.net, to search job offers). However, these search engines are developed ad hoc for each of the underlying domains, which can constitute a heavy barrier to the development of further ones.

Having into account the notion of web directories and vertical search engines, we propose to apply the disambiguation techniques previously described to group the hits retrieved by a traditional search engine into semantic categories. These categories are defined by the different meanings of the user's keywords and are used to categorize the retrieved links according to them, thus avoiding mixed-up results. Our approach, called Doctopush (Trillo et al. 2011), discovers the possible meanings of the keywords to create the categories dynamically in runtime by considering heterogeneous sources available on the Web. Differently from other proposals, such as clustering, the process of creation of categories is independent of the sources providing the results that must be shown to the users. Our approach considers two main steps: (1) discovering the semantics of the user keywords, and (2) semantics-guided data retrieval (see Fig. 4.6).

The first step, *Discovery of the Semantics of the User Keywords*, adopts the disambiguation technique previously described in QueryGen. The *Semantics-guided Data Retrieval* step pursues to provide the user with only the hits, retrieved by a traditional search engine, which she is interested in and filter out the irrelevant results. Thus, the system classifies the hits retrieved into the categories defined by the meanings of the user keywords. Moreover, the categories are ranked according to the interests of the user. In other words, the system attempts to select the hits that have the same semantics as the intended meaning of the user keywords and discard the others. This process is performed in four phases in runtime:

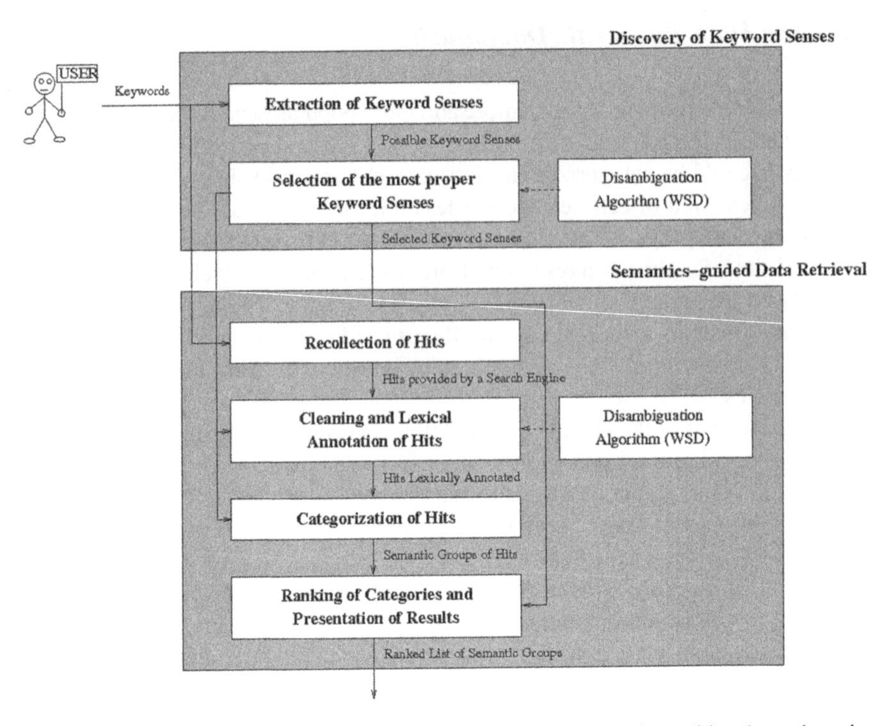

Fig. 4.6 Overview of Doctopush: A semantic prototype to group hits of a traditional search engine

- *Recollection of Hits:* The system performs a search in a traditional web search engine considering the user keywords as input. This search returns a set of relevant ranked hits, which represent the web pages where the keywords appear. The ranking of the hits depends on the specific techniques inner to the traditional search engine used (Google, Yahoo, Bing, etc.). Then, the hits returned are provided as input to the next phase (*Cleaning and Lexical Annotation of Hits*) incrementally, in *blocks of hits* of a certain size. In this way, new hits can be retrieved while the first blocks are being processed.
- *Cleaning and Lexical Annotation of Hits:* Each hit obtained in the previous phase (composed of a title, a URL, and a snippet) is automatically annotated lexically. Thus, first, each hit (H_j) goes through a *cleansing* process where stopwords are filtered out (creating the filtered hit H_j'). After that, the relevant words of the title and the snippet of each filtered hit H_j' are considered to perform its *lexical annotation.* A lexical annotation is a piece of information added to a term that refers to a semantic knowledge resource such as a dictionary, a thesaurus, or any other resource which represents a general or domain-specific ontology. So, for each filtered hit H_j', this process obtains a list of annotations denoted as $H_j' \rightarrow \{(S_1^x \rightarrow score_o, \ldots, S_1^y \rightarrow score_p), \ldots, (S_n^z \rightarrow score_q, \ldots)\}$. The list of annotations represents the senses which the user keywords are likely used within that hit.

Moreover, the score associated to each annotation indicates its reliability. This is needed because, in some situations, selecting only one sense for a user keyword in a certain hit is a difficult task even for a human due to the inherent ambiguity and polysemy. For example, a keyword can appear in the same hit with different senses (e.g., in the case of a hit corresponding to a dictionary entry), or two different senses for a keyword may have some overlapping (e.g., for the keyword *star*, the meanings are an "actor who plays a principal role" and "a performer who receives prominent billing"). Therefore, a list of annotations must be considered.
In our case, the annotation process is performed by considering each appearance of the user keywords in the filtered hit and its context (i.e., its relevant neighboring words), and by using the Web-based Word Sense Disambiguation (WSD) method and different configurations of the Probabilistic Word Sense Disambiguation (PWSD) method presented in Po (2009) and Po et al. (2009). For this, first, each appearance of a user keyword k_i in the title or the snippet of a filtered hit H_j' is marked with its probable senses, that is, $k_i^t \rightarrow \{S_i^x \rightarrow score_o, \ldots, S_i^y \rightarrow score_p\}$, where k_i^t denotes the *t*th appearance of k_i in H_j'. These annotations are used to perform the global annotation of the hit. Thus, when a user keyword sense S_i^x appears only once in the annotations performed, the sense and its corresponding score ($S_i^x \rightarrow score_o$) are incorporated to the list of annotations of the hit. Nevertheless, as a user keyword k_i could appear several times in H_j', the same user keyword sense S_i^x could appear in several annotations of k_i and have a different score in each of them; in this case, the maximum of these scores is considered for the global annotation of the hit.

- *Categorization of Hits:* The hits (already annotated as a result of the previous process) are grouped in categories by considering their lexical annotations. First, the system defines the categories that are going to be considered. Then, blocks of hits are classified. The potential categories are defined by considering all the possible combinations of candidate keyword senses of the input keywords, i.e., the Cartesian product of the candidate sense sets of the user keywords. For example, if the user introduces two keywords (k_1 and k_2) and, in the previous step, two senses are discovered for k_1 (S_1^1 and S_1^2) and one sense for k_2 (S_2^1), then the following potential categories are considered: $<S_1^1, S_2^1>$, $<S_1^2, S_2^1>$, $<U_1, S_2^1>$, $<S_1^1, U_2>$, $<S_1^2, U_2>$, and $<U_1, U_2>$, where U_1 and U_2 represent the unknown meanings considered for the keywords k_1 and k_2, respectively. Then, each hit is assigned to the categories defined by the meanings of the input keywords corresponding to the lexical annotation of that hit. So, depending on the scores of the meanings that are assigned to the user keywords in a hit, the hit could be classified in different categories at the same time, that is, the categories may overlap. Finally, the hits classified in a category are ranked according to their relevance for that category. That is, the system performs a *score-based ranking*. For this purpose, when a hit is assigned to a category, a score is also computed for the hit. This score is calculated by multiplying the scores associated to that hit for the different senses defining the category. Then, the hits within the category are ranked according to their scores (hits with the same score are ranked according to the order returned by the web search engine), as the hits in top positions are considered more relevant for that category.

- *Ranking of Categories and Presentation of Results:* Finally, the results of the Categorization of Hits phase are presented to the user. The system shows, in different tabs or category links, the categories considered that contain hits (Potential categories with no hits represent combinations of senses of the input keywords that are not detected in the hits collected). The order of the tabs or category links depends on the probability that the corresponding category represents the semantics that the user had in mind when she wrote her query. So, to rank the categories, three elements are considered: (1) the scores obtained previously, (2) the percentage of hits classified in the category, and (3) the position of the first hit in that category in the ranking provided by the web search engine. Thus, the global score for a category C_x is defined in the following way:

$$score(C_x) = \alpha score_{hitSenses} + \beta score_{\%hits} + \gamma score_{pos1stHit}$$

where α, β, and γ are the coefficients to tune the formula; $score_{hitSenses}$ is obtained by multiplying the scores (computed in previous phases) for the senses defining that category; $score_{\%hits}$ is equal to the number of hits assigned to the category C_x divided by the number of hits retrieved from the web search engine; and $score_{pos1stHit}$ is the inverse of the position of the first hit in C_x in the ranking provided by the traditional web search engine considered. Moreover, categories with unknown senses are considered less relevant, by assuming that the component $score_{hitSenses}$ is zero.

After developing the prototype, its performance under different contexts was evaluated in order to evaluate the interest of our proposal; for further details, see Trillo et al. (2011). We have also analyzed several related works in the following areas: clustering and categorization methods of documents, search engines that perform clustering of web documents, semantic search engines with the same goal as our proposal, and works about query reformulation and refinement. The main difference of our proposal with respect to other methods is that it considers the knowledge provided by ontologies available on the Web in order to dynamically define the possible categories for classifying the hits considered. Thus, it is independent of the sources providing the results that must be grouped.

4.3 Semantic Information Extraction: GENIE

The access to large amounts of information has become something regular in our daily life, and this has raised the need for more intelligent tools to collect, organize, analyze, and distribute all this information. These tools require capabilities that are not trivial and that can hardly be found in commercial products. To ease the development of such tools, it would be very useful to have off-the-self software that would comprise different elements to tackle these problems from different points of view under a common framework, providing solutions to many usual processes related to the extraction of information. This would help to increase productivity of organizations, and save resources needed to achieve their goals.

In this context, when it comes to handling information in nonstructured documents, many tasks are still open research problems. Among these tasks, we can find, for example, automatic text classification, summarization of text sources, extraction of data from raw text, or synthesis of knowledge out from natural language documents. The implementation of software to face these kinds of issues is not at all a trivial matter. Thus, in our research group, we have applied the know-how acquired developing different semantic information systems to develop a framework to help these tasks, called the GENIE project.

GENIE is the acronym for GENeric Information Extraction Framework. According to Russell and Norvig (Russell and Norvig 2003), Information Extraction (IE) means automatically retrieving certain type of information from natural language text. They say that IE is halfway between Information Retrieval (IR) systems and text understanding systems. GENIE is an architectural proposal that implements a set of components whose objective is to provide tools to make IE easier for the developers, integrating Semantic Web techniques with Machine Learning, Artificial Intelligence (AI) techniques, and Natural Language Processing (NLP) tools (Smeaton 1999). In particular, it supplies a set of libraries designed to assist developers in projects of this nature. An important feature that distinguishes this project from other similar works is the semantic approach given to all the modules, taking advantage of the latest advances made in ontological engineering and semantic technologies. To sum up, these are the main goals of GENIE:

- To create a framework able to handle different languages, and to integrate a large number of processes related to IE.
- To integrate into this framework, modular, generic, and open tools that can be used in other external applications.
- To develop an open framework allowing future expansion.
- To facilitate experimentation and testing allowing the improvement of current methods, and the development of new tools that represent an innovation in the field of IE.

The architecture of the GENIE framework is composed of a set of modules that implement essential tasks to execute different semantic IE processes. GENIE consists of a set of high-level units which can be orchestrated to form a semantic IE workflow by communicating with each other using XML. These units, if necessary, can be transformed into services, libraries, or web services, depending on the degree of decoupling and performance required. Specifically, GENIE is constituted by the following units:

- *Multilingual natural language analysis unit:* The aim of this unit is to provide basic information from raw texts. This information is similar to the information obtained by tools like morphological taggers or syntactic parsers.

- *Named Entities Detector:* This unit provides a specific semantic analyzer focused on Named Entities (NE; Sekine and Ranchod 2009), as they are subject of intense research in the context of IE. The recognition of NE has been enhanced with NLP and semantic methods which improve the term disambiguation.
- *Machine Learning unit:* This unit is another important piece inside the GENIE architecture, as it provides several unsupervised learning methods. Apart from Support Vector Machines (SVM; Joachims 1998), which have provided us very good results, this unit can also manage other unsupervised learning methods like clustering and statistical models like Bayesian networks. We have paid special attention to enhance the interconnectivity of these techniques with the rest of the units to provide a unified development framework.
- *Geographic information extractor:* Geographical resources have a special treatment due to their importance. Typically, when categorizing a text, about 30% of the labels used by a documentation department are related to places. Thus, as detailed in Garrido et al. (2013b), we have incorporated the use of ontologies to enhance the geographical tagger service provided by this unit.
- *Categorization unit:* The goal of this unit is to automatically classify documents (Garrido et al. 2011, 2012). This unit uses most of the functions supplied by the aforementioned units.
- *Semantic analyzer:* This unit obtains information about the connotation of terms, that is, the real meaning of words in a context. Moreover, it searches relations of these words with others (e.g., synonyms, antonyms, hypernyms, hyponyms, etc.) using both NLP tools and ontologies to do this task. Finally, it is also used in order to disambiguate terms.
- *Automatic query expansion unit:* This unit provides an enhanced keyword-based search by expanding automatically the input query taking into account the semantic contents of the keywords and their relationships. The approach of this GENIE's unit consists of three steps: (1) *obtain words with common lemmas*, to extract those words that belong to the same family as the keywords entered as a query; (2) *obtain words with similar meanings*, to return records which contain words that are synonymous to the keywords introduced, using the previous units to analyze and to disambiguate the terms; and finally, (3) *refine the queries with NE*, by detecting and considering NEs as a whole. Further details are provided in Granados-Buey et al. (2014a, b).
- *Aggregation unit:* This unit provides personalized reports from raw text sources. These reports are elaborated from a set of predefined templates that define the presentation of the information of some types of results (e.g., people, companies, events, etc.). The sources can be documents, databases, triplestores, or even Linked Data. Some details of this unit can be found in Garrido et al. (2014a).

Our main contribution with GENIE is not only the effort of packing these different work units but also the use of semantic techniques for tasks which are usually tackled using purely linguistic approaches or machine learning. The advantages of joining machine learning approaches with semantic tools have been widely studied in Garrido et al. (2014c). Finally, another useful feature of GENIE is that it can incorporate new resources (databases, gazetteers, dictionaries, thesauri, RDF/OWL files, etc.) to enrich the service provided by work units. This incorporation may be progressive and is performed using standard formats for resources.

Regarding other existing projects with similar goals as GENIE's, we have to mention GATE (Cunningham et al. 2002). GATE is a Java suite of tools developed at the University of Sheffield, which began in 1995 and is used nowadays by a large community of scientists, companies, teachers, and students for NLP tasks. GATE is inspired by previous projects such as ATLAS (Bird et al. 2000). GATE is, on the one hand, an integrated development environment, and, on the other hand, a framework including a set of software building blocks ready to be used. It is a mature tool, very powerful and complete, working very well with English, German, and French. However, this framework does not give the same kind of support for languages such as Spanish, Italian, or Portuguese. In any case, the main difference between GENIE and GATE is the level where each project works at. GENIE is not only a set of tools ready to be used individually but also a collection of units prepared to be connected to other systems and resources, with a level of encapsulation and coupling adaptable to the needs of each project. In fact, GENIE could integrate a suite like GATE as another working unit.

Practical Applications. In addition In addition to its high interest as a research platform, this software has a lot of practical applications:

- *Search engines:* We have increased the performance of a standard index of a term-based search engine, making its behavior closer to a semantic search one. This allows to get results in spite of the fact that gender (male or female), number (singular or plural), or verb forms are different from the keywords used in the query. It also considers synonyms and related words when retrieving data, which greatly enhances the user experience. Further details can be found in Granados-Buey et al. (2014b).
- *Documentary collections:* GENIE can help to improve the productivity of, for example, administration staff or a documentation department, by automating tedious tasks of labeling. It has tools capable of categorizing documents using semantic tags with a high-precision level. Moreover, through a suitable interface, it could also update item details in the document database when necessary. A GENIE module has been linked to geographical databases, making it capable of taking into account tagged text locations and disambiguating it when needed. Finally, GENIE is also able to produce summaries of text with a defined length, something very useful in many areas. Hypatia (Garrido et al. 2014a) is an ongoing project that brings together all of these functionalities.
- *Information management:* With the ability of "understanding" a text, GENIE can extract information from a document and transform it to measurable values. This is very interesting as it can be applied to the analysis of reports, to obtain brand reputation, or to implement filters, among others. This extracted information can be captured and represented as a knowledge model. In particular, GENIE has been used to develop tools like (Garrido et al. 2013), a system able to generate

Topic Maps (Pepper and Moore 2001), a simple form of knowledge representation, out from plain texts.

- *Recommendation systems:* Today, it is quite common for e-commerce and bookmarking web sites to include some type of recommendation module that is able to identify and present items appealing to their users. Many techniques related to areas such as machine learning, IR, or NLP, among others, have been adopted to develop systems that recommend items like books, songs, or movies, for example. Even though recommendation systems have been developed for the past two decades, existing recommenders are still inadequate to achieve their objectives and must be enhanced to generate appealing personalized recommendations effectively. In this context, we have already proposed two approaches to recommender systems, SOLE-R (Garrido et al. 2014b) and TMR (Garrido and Ilarri 2014), which exploit the semantic services provided by GENIE.

The outcomes obtained by this system on real environments are very promising, and, in fact, this framework is already being used in real production environments, providing very good results; thanks to the use of semantic techniques.

4.4 Technologies for Fuzzy Knowledge

Despite the advantages of ontologies, it has been widely pointed out that they are not appropriate to deal with imprecise and vague information, which is inherent to several real-world domains. Let us discuss now some examples. The domain of medicine contains a lot of imprecise terms, and classical ontologies are not suitable to express that a patient is *slightly* unconscious or that anaphylaxis is *quite similar* to sepsis. Location-based applications are based on potentially imprecise sensor data: For instance, GPS sensors provide an approximate location. The domain of accommodations includes vague terms to categorize different lodging types (such as *Guesthouse*). For example, it is usually assumed that a *Guesthouse* is a "cheap, small, and more hospitable hotel", but the notions of *cheap, small*, and *hospitable* are clearly not well defined, having unsharp boundaries. Hence, the nature of the concepts in this domain makes crisp definitions unsuitable. For more applications of fuzzy ontologies, see Bobillo (2008).

Fuzzy set theory and fuzzy logic have proved to be suitable formalisms to handle this imprecise and vague knowledge (Zadeh 1965). In fuzzy logic, the usual convention prescribing that a statement is either true or false is changed. Every statement holds with a *degree of truth* measured on an ordered scale that is no longer $\{0, 1\}$ but (usually) $[0, 1]$. The main concept in fuzzy logic is that of fuzzy set. Essentially, the elements of a fuzzy set have degrees of membership valued in $[0, 1]$. For example, Fig. 4.7 shows a fuzzy set representing young people.

Fuzzy logics provide compositional calculi of degrees of truth. The conjunction, disjunction, complement, and implication operations are performed in the fuzzy case by a t-norm function $\otimes$, a t-conorm function $\oplus$, a negation function $\ominus$, and an implication function $\Rightarrow$, respectively. A quadruple composed by a t-norm, a t-

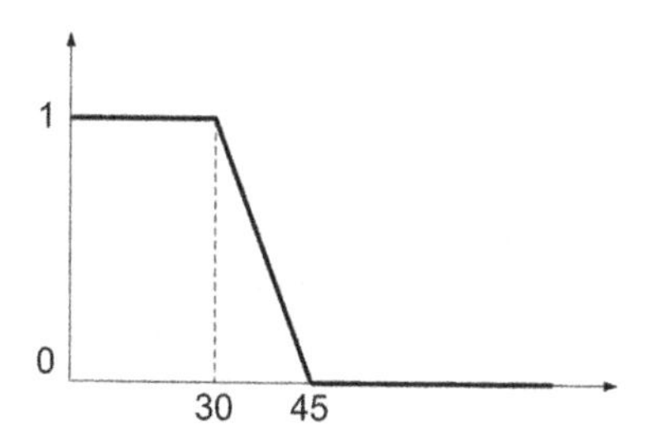

Fig. 4.7 Fuzzy set representing young people

conorm, an implication function, and a negation function determines a fuzzy logic. Table 4.3 shows the definition of some popular fuzzy logics, where $\alpha,\beta \in [0,1]$ are two degrees.

Fuzzy Ontologies Given the success and popularity of fuzzy logic, it should come as no surprise that several fuzzy extensions of ontologies can be found in the literature. The elements of a fuzzy ontology are extended as follows:

- *Fuzzy concepts* are interpreted as fuzzy sets of individuals, so the membership of an instance to a concept is a matter of degree. For example, *YoungHuman* can contain the fuzzy set of young people.
- Similarly, *fuzzy properties* between two individuals are interpreted as fuzzy relations, so pairs of elements are related to some degree. For instance, the property *isFriendOf* makes it possible to represent the degree of friendship between pairs of individuals.
- Now, it makes sense to consider *fuzzy axioms,* since statements are not either true or false but hold to some degree. For example, we can state that *john* belongs to the concept of *YoungHuman* with at least degree 0.9, meaning that he is rather young. Classical axioms τ are generalized as (τ,α).
- Finally, it makes sense to consider *fuzzy datatypes* generalizing crisp values by using a fuzzy membership function. For example, instead of considering the crisp value 18, now it is possible to consider *about18*. The former datatype is incompatible with the value 17.99, whereas the latter one is not.

Table 4.3 Truth combination functions of various fuzzy logics

Operator	Lukasiewicz logic	Gödel logic	Product logic	Zadeh logic
$\alpha \otimes \beta$	$\max(\alpha+\beta-1, 0)$	$\min(\alpha, \beta)$	$\alpha \cdot \beta$	$\min(\alpha, \beta)$
$\alpha \oplus \beta$	$\min(\alpha+\beta, 1)$	$\max(\alpha, \beta)$	$\alpha+\beta-\alpha \cdot \beta$	$\max(\alpha, \beta)$
$\alpha \Rightarrow \beta$	$\min(1-\alpha+\beta, 1)$	$\begin{cases} 1 \text{ if } \alpha \leq \beta \\ \beta \text{ otherwise} \end{cases}$	$\min(1, \beta/\alpha)$	$\max(1-\alpha, \beta)$
$\ominus\, \alpha$	$1-\alpha$	$\begin{cases} 1 \text{ if } \alpha = 0 \\ 0 \text{ otherwise} \end{cases}$	$\begin{cases} 1 \text{ if } \alpha = 0 \\ 0 \text{ otherwise} \end{cases}$	$1-\alpha$

The reasoning services in fuzzy ontologies include the same tasks for classical ontologies (see Section 4.1) together with some new ones. The most typical tasks are:

- *Consistency:* An ontology O is consistent iff it has a model, i.e., there is a fuzzy interpretation satisfying every axiom in O.
- *Entailment:* O entails an axiom τ iff every model of O satisfies τ.
- *Concept satisfiability:* A concept C is satisfiable to at least degree α (or α-satisfiable) w.r.t. O iff there is a model of O satisfying $(C(x),\alpha)$, for a new individual $x \notin O$.
- *Concept subsumption:* A concept D subsumes a concept C to at least degree α (or α-subsumes) w.r.t. O iff every model of O satisfies the axiom $(C \sqsubseteq D,\alpha)$.
- *Best entailment degree* (*BED*): The BED of a crisp axiom τ w.r.t. O is defined as the supremum of the degrees α such that O entails the axiom (τ,α), where $sup\,\varnothing = 0$.

Fuzzy ontologies require the development of new languages, reasoning algorithms, and tools. Unfortunately, they have not achieved yet the maturity of crisp ontologies, and additional research on this topic is still being carried out. However, there exist several approaches in the literature and implementations worth to mention. This section summarizes the main achievements developed by members of our research group in collaboration with international experts.

There is no standard language to represent fuzzy ontologies, and the different proposals have differences in the elements that are being fuzzified. For example, not all the approaches consider the definition of fuzzy sets using trapezoidal membership functions. To assist users in the process of fuzzy ontology building, there is a Protégé plug-in called *Fuzzy OWL 2* (http://webdiis.unizar.es/~fbobillo/fuzzyOWL2) that can be used to create and edit fuzzy ontologies (Bobillo and Straccia 2011). The plug-in uses a relatively abstract fuzzy ontology representation that can be exported into the particular syntax of different ontology languages.

Among the many existing implementations of fuzzy ontology reasoners, we will highlight two: *fuzzyDL* (http://webdiis.unizar.es/~fbobillo/fuzzyDL; Bobillo and Straccia 2008) and *DeLorean* (http://webdiis.unizar.es/~fbobillo/delorean; Bobillo et al. 2012). The former system implements a fuzzy extension of a classical tableaux algorithm. The latter follows an alternative approach and transforms a fuzzy ontology into an equivalent non-fuzzy ontology, preserving the semantics of the knowledge in such a way that it is possible to reuse the existing ontology reasoners. All the existing fuzzy ontology reasoners are complementary, because up to now, they support different fuzzy ontology elements, and hence cannot be easily ranked.

The rest of this section is dedicated to a deeper overview of these three applications: Fuzzy OWL 2, fuzzyDL, and DeLorean.

Modeling Fuzzy Ontologies with Fuzzy OWL 2

Fuzzy OWL 2 is a Protégé plug-in that makes it possible to develop fuzzy ontologies. Once the plug-in is installed, a new tab in Protégé, named Fuzzy OWL, enables to use it.

The plug-in is based on a methodology for fuzzy ontology representation using OWL 2 (Bobillo and Straccia 2011). The key idea of this representation is to use an OWL 2 ontology, and extend its elements with annotations representing the features of the fuzzy ontology that OWL 2 cannot directly encode. To separate the annotations including fuzzy information from other annotations, a new annotation property called *fuzzyLabel* is used, and every annotation is identified by the tag *fuzzyOwl2*. Since typing such annotations is a tedious and error-prone task, the plug-in makes the syntax of the annotations transparent to the users.

Figure 4.8 shows the available options of the plug-in: fuzzy queries, fuzzy ontologies, fuzzy modifiers, fuzzy concepts (fuzzy nominals, fuzzy-modified concepts,

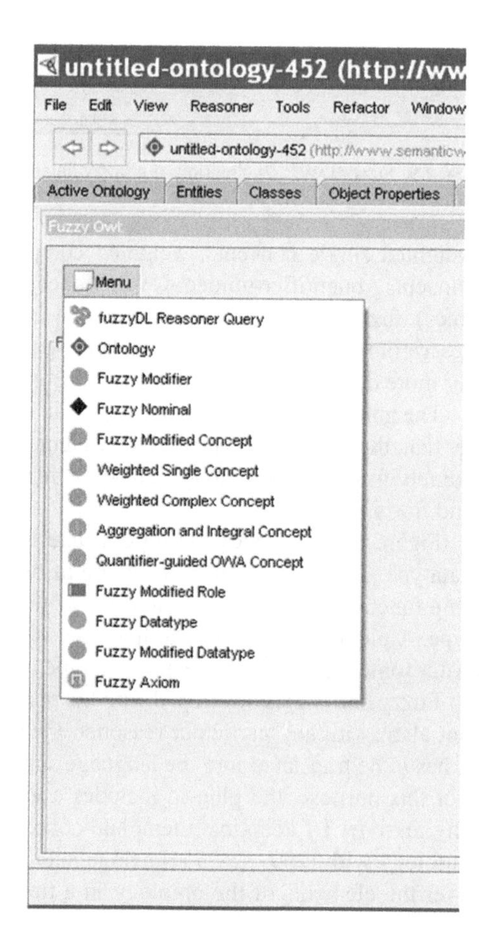

Fig. 4.8 Fuzzy OWL 2: Menu options

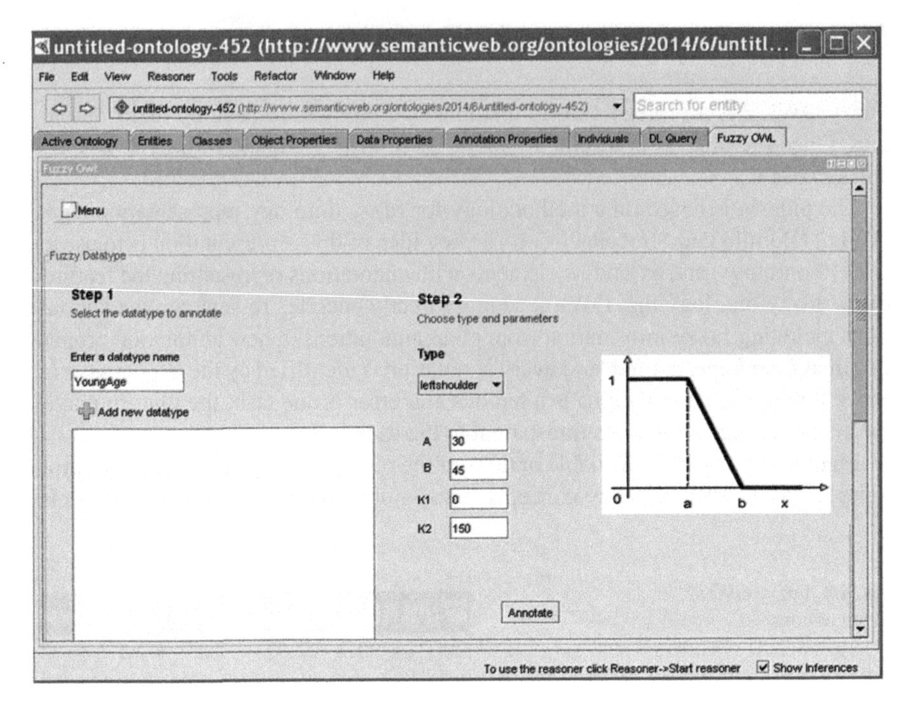

Fig. 4.9 Fuzzy OWL 2: Creation of a fuzzy datatype

weighted single concepts, weighted complex concepts, aggregation and integral concepts, quantifier-guided OWA concepts), fuzzy properties (fuzzy-modified roles), fuzzy datatypes (including fuzzy-modified datatypes), and fuzzy axioms. A description of all the elements of a fuzzy ontology is out of the scope of this chapter; for more details, we refer the reader to Bobillo and Straccia (2011, 2013).

The non-fuzzy part of the ontology can be created by using Protégé as usual. After that, the user can define the fuzzy elements of the ontology by using the plug-in, namely fuzzy axioms, fuzzy datatypes, fuzzy modifiers, fuzzy-modified concepts, and fuzzy nominals.

Figure 4.9 illustrates the plug-in use by showing how to create a new fuzzy datatype. The user specifies the name of the datatype and the type of the membership function. Then, the plug-in asks for the necessary parameters according to the type. A picture is displayed to help the user recall the meaning of the parameters. After some error checks, the new datatype is created and can be used in the ontology.

Fuzzy ontologies developed with the plug-in use an independent syntax that does not align with any particular reasoner. Once the fuzzy ontology has been created, it has to be translated into the language supported by a specific fuzzy DL reasoner. For this purpose, the plug-in includes a general parser that can be customized to any reasoner by adapting a template code. The parser browses the contents of the ontology with *OWL API 3* (Horridge and Bechhofer 2011), which allows iterating over the elements of the ontology in a transparent way and prints an informative message. The template code has been adapted to build two parsers, one for fuzzyDL and one for DeLorean. Furthermore, similar parsers for other fuzzy DL reasoners can be easily obtained. To do so, one can just replace the default messages by well-formed axioms, according to the desired fuzzy ontology syntax.

Reasoning with Fuzzy Ontologies Using fuzzyDL

From a historical point of view, *fuzzyDL* can be considered as the first fuzzy DL reasoner. It is very popular and has been used in several applications. The supported language is thus a fuzzy extension of $\mathcal{SHIF}(\mathbf{D})$. It is also possible to use linguistic labels as degrees of truth, such as *high*, instead of forcing fuzzy ontology developers to use numbers in [0, 1]. This makes it possible to deal with unknown degrees of truth.

Apart from extending the elements of crisp ontologies to the fuzzy case, fuzzyDL introduces new elements specific from the fuzzy case, such as concept modifiers (using linear hedges and triangular functions), explicit definitions of fuzzy concepts (by means of triangular, trapezoidal, left-shoulder and right-shoulder functions), or some concept constructors (fuzzy rough concepts (Bobillo and Straccia 2012), aggregation operators, modified concepts, and threshold concepts). It is also possible to express linear inequations involving degrees of truth. The semantics is given by Zadeh and Lukasiewicz fuzzy logics, and some operators of Gödel fuzzy logic are also supported. Connectives of different fuzzy logics can be combined.

fuzzyDL supports standard reasoning tasks namely, consistency, concept satisfiability, maximum degree of satisfiability of a concept, entailment, concept subsumption, BED of an axiom, maximum degree of satisfiability of an axiom, and instance retrieval. Furthermore, it also supports some nonstandard tasks, such as variable optimization and different types of defuzzification of fuzzy sets.

The reasoning algorithm is based on a mixture of tableau rules and an optimization problem. All the reasoning tasks are reduced to the BED of a concept assertion. Then, it applies some satisfiability preserving tableau rules that not only decompose complex concept expressions into simpler ones but also generate a system of inequation constraints. These inequations have to hold in order to respect the semantics of the fuzzy DL constructors. After all the rules have been applied, an optimization problem must be solved before obtaining the final solution.

fuzzyDL implements several optimization techniques, such as general concept inclusion (GCI) absorption, concept simplification, lexical normalization, optimized use of n-ary conjunction and disjunction, advanced blocking techniques, normalization of degrees of truth, encoding of string names using integers, etc.

Although fuzzyDL can be freely downloaded, the user needs a *Gurobi* solver license because it is required to calculate the solutions of mixed integer linear programming (MILP) problems. fuzzyDL can be used as a stand-alone application, accessed by other applications by means of a Java API, or queried through the Fuzzy OWL 2 plug-in.

Reasoning with Fuzzy Ontologies Using DeLorean

DeLorean (DEscription LOgic REasoner with vAgueNess) is a fuzzy ontology reasoner that supports fuzzy extensions of the languages OWL and OWL 2. Nowadays, DeLorean is the only reasoner that currently supports fuzzy OWL 2, although it does not support several elements supported by other fuzzy DL reasoners.

The reasoning algorithm is based on a reduction to reasoning in crisp ontologies detailed in Bobillo et al. (2012b). A consequence of the reduction is the possibility to reuse classical resources: editors, tools, reasoners, etc. In a strict sense, DeLorean is not a reasoner but a translator from a fuzzy rough ontology language into a classical ontology language (the standard language OWL or OWL 2, depending on the expressivity of the original ontology). Then, a classical DL reasoner is employed to reason with the resulting ontology. Nevertheless, due to this ability of combining the reduction procedure with the classical DL reasoning, we refer to DeLorean as a reasoner.

The supported language is thus a fuzzy rough extension of $\mathcal{SROIQ}(\mathbf{D})$. It is also possible to use linguistic labels as degrees of truth. The semantics is given by Zadeh and Gödel fuzzy logics, and connectives of both logics can be combined. The following reasoning tasks are supported: Computation of an equivalent crisp representation of the fuzzy ontology, consistency, concept satisfiability, maximum degree of satisfiability of a concept, entailment, concept subsumption, and BED of an axiom. There are some optimizations of the reduction to crisp $\mathcal{SROIQ}(\mathbf{D})$, but the only existing optimizations for reasoning in crisp $\mathcal{SROIQ}(\mathbf{D})$ are those implemented by the reused classical reasoner.

DeLorean can be used as a stand-alone application. In addition, DeLorean reasoning services can also be used from other programs by means of a Java API. In this section, we focus on the use of the reasoner through its graphical interface, which as illustrated in Fig. 4.10, is structured in four sections:

- *Input*. Here, the user can specify the input fuzzy ontology and the DL reasoner that will be used in the reasoning. The possible choices are HermiT (Motik et al. 2009), Pellet (Sirin et al. 2007), and any OWLlink-complaint reasoner. Once a fuzzy ontology is loaded, the reasoner will check that every degree of truth that appears in it belongs to the set specified in the section on the right.
- *Degrees of truth*. The user can specify here the set of degrees of truth that will be considered. 0 (false) and 1 (true) are mandatory. Other degrees can be added, ordered (by moving them up or down), and removed. For the user's convenience, it is possible to directly specify a number of degrees of truth, and they will be automatically generated.
- *Output*. Here, output messages are displayed. Some information about the reasoning is shown here, such as the time taken, or the result of the reasoning.
- *Reasoning*. This part is used to perform the different reasoning tasks that DeLorean supports. The panel is divided into five tabs, each of them dedicated to a specific reasoning task.

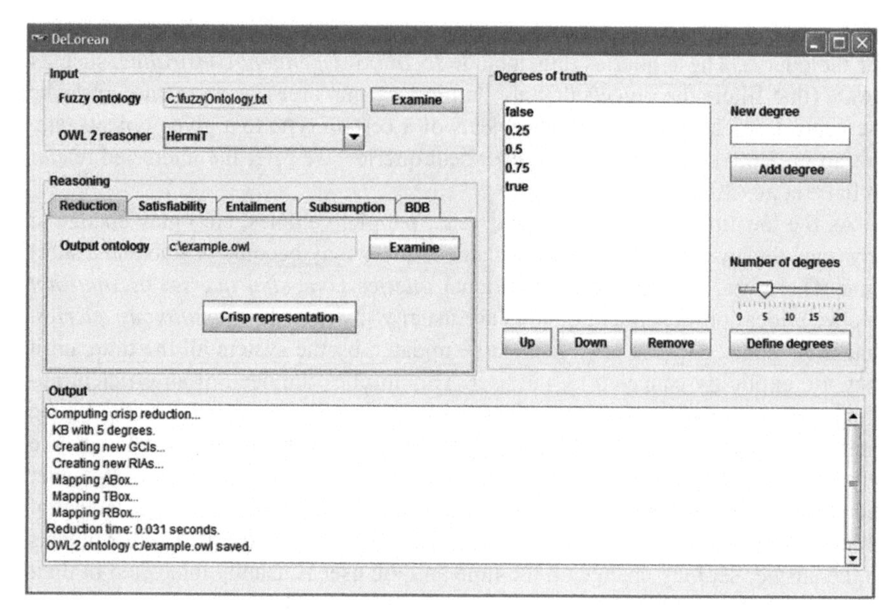

Fig. 4.10 User interface of DeLorean reasoner

These three technologies show how the expressivity of crisp DL ontologies can be extended using a complementary formalism. In our case, we focused on fuzzy logic, as it provided us with mechanisms to model the uncertainty that is inherent in many of our working scenarios.

4.5 Applying Semantic Web Technologies to Mobile Computing

Advances in mobile computing, with the popularity and widespread and pervasive use of mobile devices and wireless communication technologies, have emphasized the interest in providing mobile users with access to useful information, anywhere and anytime. Besides, in this type of scenarios, it is particularly important to customize the information provided according to the context of the user, to show her only truly relevant data and avoid overloading the user with unneeded information. One of the most important context parameters is the location of the user, which has given rise to the development of a wide variety of LBSs (Ilarri et al. 2010; Schiller and Voisard 2004), such as vehicle tracking applications, friend-finder applications, location-based emergency services, location-based advertisements, and location-based games, among others.

A fundamental building block to define appropriate LBSs for a variety of scenarios is the so-called *location-dependent queries*, which are queries whose answer depends on the location of certain objects/entities (notably, not only the location of the mobile user that submits the query but also other objects that are relevant for the query). These queries thus include *location-dependent constraints*, such as *inside* (that filters the objects that are within a certain area), *nearest* (that retrieves the nearest object or the k-nearest objects of a certain type to a given object), etc. For an in-depth study of location-dependent queries, we refer the interested reader to Ilarri et al. (2010).

As the locations of moving objects (e.g., people, vehicles, etc.) may change at any time, the answer to a location-dependent query may become obsolete in a short time. Therefore, as opposed to traditional queries (*one-shot queries* or *snapshot queries*), location-dependent queries are usually processed as *continuous queries*, that is, as queries whose answer must be updated by the system all the time, until they are explicitly canceled by the user. This implies the need of an efficient approach to keep the answer up-to-date, without the need to reevaluate the whole query from scratch every time that the answer needs to be refreshed. Besides, the answer to a query is usually refreshed periodically: Triggering a query whenever the answer changes is usually not possible, as even if the answer set (i.e., the set of objects that are an answer to the query) does not change, the locations of the objects in the answer set may change all the time and the user is usually interested in their current locations, that therefore will need to be refreshed continually.

It is important to highlight that sometimes a GPS location granularity for location-dependent queries may be unnecessary or even inconvenient. So, the user should be allowed to express locations and location constraints in terms of the location terminology that she requires (the locations could be not only GPS locations but also cities, rooms, buildings, provinces, countries, etc.), which are called *location granules* in Ilarri et al. (2011a). Its use may have an impact on the semantics of the query, on the performance of the query processing, and also on the way the query results are shown to the user. Moreover, location granules should not be considered just a set of geographic locations, as they may also have other features and additional meaning attached; for this reason, we have later proposed the concept of *semantic location granule* (to be described in the next section).

Location-dependent queries would benefit from the semantic management of information about moving objects and scenarios. Acquiring the required knowledge about the existing types of objects and their features, the application scenarios, etc., should not be the responsibility of the user. This has motivated our ongoing work on the SHERLOCK system (to be described later), where we also try to go beyond simple location-dependent queries to more generic *location-dependent requests*, which may be traditional location-dependent queries but also commands requesting objects to perform specific actions (e.g., using their sensors to measure certain values).

Semantic Location Granules

As we have mentioned in the previous section, location-dependent queries are one of the most active matters of study in LBSs. For example, if we are visiting New York and take with us a smartphone, we could query "what are the museums in 1 kilometer?" In the literature, most of the works (Ding et al. 2008; Gedik and Liu 2006; Ilarri et al. 2010; Mokbel et al. 2005; Sistla et al. 1997) consider that the answer to this query is a set of GPS points (that could be located on a map). But in many cases, we do not need the precise GPS location. Instead, we need to use a more abstract notion of location, *area name*, since the geographic coordinates are probably meaningless to us. For instance, if a user wants to go by subway to the MoMA museum in New York, she will not be interested in the precise coordinates of the area where she has to leave the subway, but she may want to know the name of the station near the MoMA museum to leave the subway. It makes no sense to show her a map with a point to leave the subway. She needs the name of the station. To take these situations into account, the concept of *location granule* arises as a set of physical locations (Bernad et al. 2013; Bobed et al. 2011; Ilarri et al. 2007, 2009, 2011). This concept is similar to the concept of *place* in Hightower (2003) and Hoareau and Satoh (2007, 2009) or *spatial granule* in Belussi et al. (2009). Furthermore, we can group a set of location granules under a name to obtain what it is known as a *location granules map*. For example, Madrid could be a location granule of the location granules map of the provinces of Spain.

Although location granules enhance the expressivity of location-dependent queries and this is an important step forward, they are basically a set of GPS points with a name. When we group a set of locations and give them a name, we are implicitly giving them also a meaning. For example, the set of locations that compose Madrid, that is, the location granule with the name Madrid, becomes a city, the capital of Spain. Thus, the location granule Madrid can be seen as a more abstract concept that represents a city or a capital of a country. Implicitly, a location granule has a semantics (it is a city, a country, etc.). To model the semantics of location granules, the concept of *semantic location granule* is introduced in Bernad et al. (2013) and Bobed et al. (2010).

In addition, it is also interesting to consider not only the semantics of the location granules (Madrid is a city, Hyde Park is a park, Spain is a country) but also the semantics of the topological relations between them: For instance, Madrid *is contained* in Spain, or Oxford Street *is adjacent* to Hyde Park. And what is more important is that, if we introduce semantics in location granules and in its topological relations, we can infer implicit knowledge automatically. For instance, if a user is in the city of Zaragoza, it could be inferred that she is also close to France, a country with famous red wines. The interest of linking LBSs and semantics has been emphasized in Ilarri et al. (2011b).

Several works in the field of geographic information systems (GIS; Rigaux et al. 2002; Shekhar and Chawla 2002) have used the relational data model to represent topological relations (e.g., the region connection calculus, RCC; Randell et al. 1992). However, to support reasoning with geographic elements, several previous works have studied the introduction of ontologies in the area of GIS (Couclelis 2010; Lutz and Klien 2006), and the introduction of different types of topological relations in DLs (Haarslev et al. 1998; Lutz and Möller 1997). Despite these efforts, more research is needed in this area to effectively enhance the processing of location-dependent queries with inference capabilities over location granules. For example, *isContained* is a key topological relationship (it allows representing the geographic hierarchy of areas), but the existing proposals to represent such a relationship have important disadvantages (depending on the specific proposal, wrong conclusions may be obtained or a manual assertion of many of the *isContained* relations may be required).

Summing up, the three main goals that are pursued with semantic location granules are:

- To represent the semantics of a set of locations, i.e., to represent the semantics behind location granules.
- To represent the semantics of topological relations between location granules.
- To be able to infer implicit knowledge automatically.

We have proposed two complementary models for semantic location granules (Bobed et al. 2010; Bernad et al. 2013). The former one considers the location granules as instances of a concept *Granule*, and it is oriented to exploit the ABox extending the query model using logical rules; for further details, see Bobed et al. (2010). In this chapter, we will focus on the latter one, which considers that the granules themselves are concepts, as they subsume a set of locations (which now become the instances). As we will see, this latter model allows the DL reasoner to make intensive use of the TBox to infer the containment relationships (Bernad et al. 2013). In the following, we will say granule instead of location granule, for the sake of simplicity.

Modeling Semantic Location Granules as Concepts Let us now discuss modeling semantic location granules as concepts. In this approach, we will consider that a semantic location granule is a concept in a TBox $\mathcal{T}$ of a knowledge representation $\mathcal{K}=(\mathcal{T},\mathcal{A})$. For example, we will consider that *Madrid* and *Spain* are concepts in $\mathcal{T}$. The most straightforward way to express that Madrid is contained in Spain is to add an axiom in $\mathcal{T}$, *Madrid* $\sqsubseteq$ *Spain*, but it fails as we will see in the following. Recall that one of the objectives to introduce semantic location granules is the possibility to express that Madrid is a city or Spain is a country. Again, the most natural way to express these situations is to consider the concepts *City* and *Country,* and add in $\mathcal{T}$ the axioms *Madrid* $\sqsubseteq$ *City* and *Spain* $\sqsubseteq$ *Country*. But if we want to say that a city cannot be a country, that is, *City* $\sqsubseteq \neg$ *Country*, then the T Box $\mathcal{T}$ becomes inconsistent: It is inferred that Madrid is a city and a country, *Madrid* $\sqsubseteq$ *Spain* $\sqsubseteq$ *Country*, which is a contradiction. With this simple example, we can see that it is not so trivial to express the content topological relationship using a DL. The main problem is that the subsumption operator ($\sqsubseteq$) is used to express that Madrid is an area inside the area of Spain, and to express that Madrid is a city, that is, the subsumption operator is used to express the content relationship between areas as well as the *is a* relationship between concepts.

To avoid the above problem, in the formalization of semantic granules and semantic granule maps with DLs, we use a transitive role, named *isContained*, and roles $loc_{X_1},\ldots,loc_{X_n}$. Intuitively, the role *isContained* is used to express that a granule is geographically contained in another one by subsumption and participation in the relationship *isContained* (e.g., *NewYork* $\sqsubseteq\exists$ *isContained. EEUU*); and $loc_{X_1},\ldots,loc_{X_n}$ are the coordinates of a point. These concrete features allow us to define areas; for example, $loc_x \geq 10 \sqcap loc_x \leq 20 \sqcap loc_y \geq 10 \sqcap loc_y \leq 30$ represents a rectangle.

Definition 1 An area concept $f(loc_{X_1},\ldots,loc_{X_n})$ is a concept built with the constructors $\sqcup$ and $\sqcap$, and the roles $loc_{X_1},\ldots,loc_{X_n}$. An area concept name A is a concept name such that $A \equiv f(loc_{X_1},\ldots,loc_{X_n})$, where f is an area concept. The set of names of area concepts is denoted by N_A.

For example, $\exists loc_x \geq 10 \sqcap \exists loc_x \leq 20 \sqcap \exists loc_y \geq 10 \sqcap \exists loc_y \leq 30$ is an area concept, while $\exists loc_x \geq 10 \sqcap \exists loc_x \leq 20 \sqcap \exists loc_y \geq 10 \sqcap \exists loc_y \leq 30 \sqcap City$ is not.

Now, we formalize the definition of semantic location granule and semantic granules map.

Definition 2 *Given $\mathcal{K}=(\mathcal{T}, \mathcal{A})$ a knowledge representation and a granule map $M = \{G_1,\ldots,G_n\}$ with G_i granules, a semantic granule map is a tuple (M, $\mathcal{K}$, area, semGranule) where area and semGranule are functions from the set of granules of M to the concept names of $\mathcal{T}$; that is, area, semGranule $M_G \rightarrow N_C$, such that the following must be satisfied for all $G \in M_G$.*

1. $semGranule\ (G) \sqsubseteq area\ (G)$
2. $area\ (G) \sqsubseteq \exists\ isContained.semGranule(G)$

The concept names $semGranules(G)$ are called semantic granules.

Let us show an example to explain the definition. Let M be a granule map with location granules $M=\{ZaragozaGr, AragonGr, MadridGr, SpainGr\}$, and $\mathcal{T}$ the TBox of a knowledge representation $\mathcal{K}$ defined in Table 4.4 (to keep explanations easier to follow, we represent geographic areas in the TBox by simple rectangles instead of the real geographic limits).

We define a semantic granule map (*M, $\mathcal{K}$, area, semGranule*), where *area* and *semGranule* are functions *area* (*SpainGr*)=*SpainArea*, etc., and *semGranule* (*SpainGr*)=*Spain*, etc. We can ensure that this is a semantic granule map since it holds the conditions (1) and (2) of Definition 2 from axioms (5)–(8), and (9)–(12), respectively. Figure 4.11 shows the map corresponding to the modeled area.

Intuitively, the condition (1) of Definition 2 says that a semantic granule is not only its geographic area but it could also have more attributes. For example, *Zaragoza* is an area and a *City*, and *Spain* is an area and a *Country*. The condition (2) allows to establish qualitative relations between granules such as "Zaragoza is a city in Spain," i.e., *Zaragoza* $\sqsubseteq \exists$ *isContained.Spain*, or to express the concept "Aragon's wines," *AragonWine* $\equiv$ *Wine* $\sqsubseteq \exists$ *isContained.Aragon*. Note that we do not express the concept Aragon's wine as *Wine* $\sqcap$ *Aragon*, since *Aragon* is a *Region* and wines are not regions; and similarly, Aragon's wines are not defined as *Wine* $\sqcap$ *AragonArea*, as wines could not have location information. We have

Table 4.4 TBox axioms involving semantic granules

Axiom	Definition
(1)	$ZaragozaArea \equiv \exists loc_x \leq 25 \sqcap \exists loc_x \leq 30 \sqcap \exists loc_y \geq 23 \sqcap \exists loc_y \leq 30$
(2)	$AragonArea \equiv \exists loc_x \geq 25 \sqcap \exists loc_x \leq 30 \sqcap \exists loc_y \geq 20 \sqcap \exists loc_y \leq 32$
(3)	$MadridArea \equiv loc_x \geq 15 \sqcap \exists loc_x \leq 20 \sqcap \exists loc_y \geq 17 \sqcap \exists loc_y \leq 23$
(4)	$SpainArea \equiv \exists loc_x \geq 5 \sqcap \exists loc_x \leq 35 \sqcap \exists loc_y \geq 0 \sqcap \exists loc_y \leq 35$
(5)	$Zaragoza \equiv ZaragozaArea \sqcap City$
(6)	$Aragon \equiv AragonArea \sqcap Region$
(7)	$Madrid \equiv MadridArea \sqcap City$
(8)	$Spain \equiv SpainArea \sqcap Country$
(9)	$ZaragozaArea \sqsubseteq \exists\ isContained.Zaragoza$
(10)	$AragonArea \sqsubseteq \exists\ isContained.Aragon$
(11)	$MadridArea \sqsubseteq \exists\ isContained.Madrid$
(12)	$SpainArea \sqsubseteq \exists\ isContained.Spain$
(13)	*Region, Country, City* are mutually disjoint
(14)	$RedWine \sqsubseteq Wine$
(15)	$AragonWine \equiv Wine \sqcap \exists\ isContained.Aragon$

Fig. 4.11 Sample granule map

divided the containment relationship in two parts: One to make calculations about areas (quantitative reasoning) using the subsumption relationship, and another one to establish relationships with other concepts (qualitative reasoning) using the *is-Contained* relationship.

From this model, a DL reasoner can deduce a number of facts, as we explain in the following:

Proposition 1 *A reasoner under conditions of Definition 2 can infer that:*

1. *A granule G is contained in a granule G', i.e., it can be deduced that* $semGranule(G) \sqsubseteq \exists\ isContained.semGranule(G')$
2. *A granule G intersects a granule G'*

In the example above, it can be inferred that Zaragoza is contained in Spain, $Zaragoza \sqsubseteq ZaragozaArea \sqsubseteq SpainArea \sqsubseteq \exists\ isContained.Spain$. The second statement is obvious since it is equivalent to asking if the area concept $area\ (G) \sqcap area\ (G')$ is satisfiable.

Another interesting remark is that the content of a granule only depends on its area. What does this mean? For example, let us suppose that we define two different semantic granules with the same area, *VaticanCity* and *VaticanCountry*, defined as $VaticanArea \sqcap City$ and $VaticanArea \sqcap Country$, respectively. We would like the content of Vatican as a city ($\exists\ isContained.\ VaticanCity$) to be equal to the content of the Vatican as a country ($\exists\ isContained.\ VaticanCountry$), even when a country is not a city. Due to the conditions (1) and (2) of Definition 2 and the transitivity of the role *isContained,* we can conclude that in our model $\exists\ isContained.VaticanCity \equiv \exists\ isContained.VaticanCountry$ as it is shown in the following proposition.

Proposition 2 Let *G be a granule under conditions of Definition 2. Then it holds that:*
$\exists\ isContained.semGranule\ (G) \equiv \exists\ isContained.area(G)$

and therefore, *if* G_1 *and* G_2 *are granules such that* $area\ (G_1) \equiv area\ (G_2)$, *then* $\exists\ isContained.semGranule\ (G_1) \equiv \exists\ isContained.semGranule\ (G_2)$

For further details on the model and the proofs of the different propositions, we refer the interested reader to Bernad et al. (2013).

Semantic Management of LBSs: SHERLOCK

The astonishing penetration of mobile computing in our daily lives, thanks to devices such as smartphones and tablets, leads us to a scenario where mobile users have access to huge amounts of information *anytime* and *anywhere*. Thousands of applications (also known as *apps*) for their smart devices are available to offer them information about transportation, entertainment, culture, etc. The Web has also been growing steadily in the last few years with tons of potentially useful information. Therefore, users are starting to be overwhelmed with the amount of data they receive from different sources, as it is sometimes difficult for people to distinguish which information is valuable.

For example, imagine a researcher attending a conference who arrives on an evening flight and needs to reach her city hotel. At first, she would be interested in transport information and she might need to know the different options (e.g., buses, metros, taxis, or car rental options), traffic conditions, and perhaps even where available parking spaces are located. This information could be obtained by visiting a tourist office, searching a local transportation web site, or even downloading a mobile app. After checking in, she could be interested in finding other nearby conference attendees to talk to them or even to go sightseeing (again, she should browse the Web to find information about interesting places to visit). Thus, it is the user herself who is in charge of knowing/finding the interesting and updated information sources and gathering and correlating all this information; even worse, she will have to know/find all these *updated* information about each city she visits.

Semantic Web techniques become particularly useful in these scenarios. First, they can be used to understand the information needs of users controlling the ambiguities of natural language (as already explained in this chapter). Also, these techniques can help to find the most appropriate information from a range of different sources by inferring useful information providers. Finally, information extracted from heterogeneous sources can be presented in an integrated way by using common representation models such as ontologies.

Overview We present SHERLOCK (System for Heterogeneous mobilE Requests by Leveraging Ontological and Contextual Knowledge) (Yus et al. 2014a), a system to provide mobile users with interesting LBSs. SHERLOCK (http://sid.cps.unizar.es/SHERLOCK) relieves users from the need to obtain up-to-date information about the services they need. Using ontologies to model the knowledge related to these services, SHERLOCK devices exchange information among themselves, for example, about LBSs in the area. Also, with the help of a semantic reasoner, our system is able to determine which information could be interesting for a user regarding her context, and to obtain this information from objects nearby by leveraging the collaboration among devices. For that purpose, SHERLOCK deploys a network or mobile agents (Lange and Oshima 1999; Spyrou et al. 2004; Trillo et al. 2007b) which move from one device to another autonomously to be near the needed information source and collect data (see Fig. 4.12).

Obtaining Knowledge from Devices Around SHERLOCK is based on knowledge sharing among devices. Each participating device starts with a basic OWL local ontology containing the basic terms to define LBSs, with concepts such as "Service," "Provider," "Parameter" (see Fig. 4.13 where the basic terms are in bold font). SHERLOCK devices learn from the interaction with others: When two devices meet, they share part of their local ontologies. A SHERLOCK device that receives new knowledge integrates it with its local ontology, and thus it can use this new knowledge to provide the user with more interesting information. For example, the device of a user that lives in Zaragoza (Spain) knows transportation concepts such as "Taxi" or ""Bus"; if the user travels to Thrissur (India) and her device starts

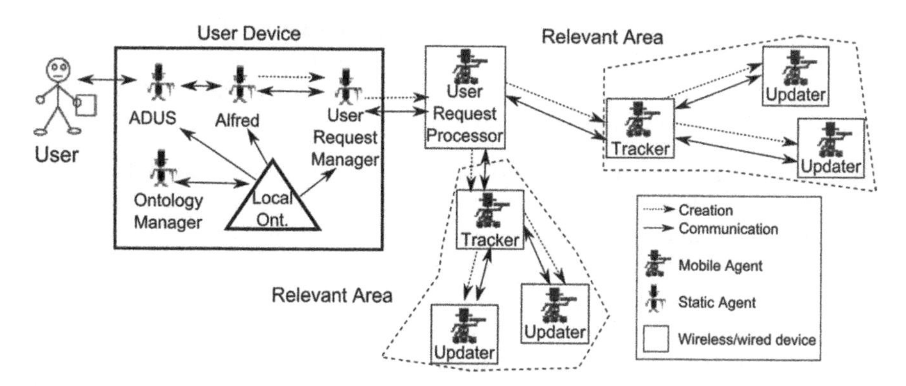

Fig. 4.12 Agent network deployed to process a request

communicating with others, it can learn that "Tuk-Tuk" (a vehicle defined in the ontology as a private transport that carries people to a certain destination) is similar to a taxi, and thus, it can be interesting for a user that needs transport.

In SHERLOCK, there is an (static) agent in charge of managing the local ontology of the device. This agent, called *Ontology Manager* (OM), has two main tasks: (1) sharing knowledge with OMs in other devices, and (2) integrating the knowledge received. OMs are continuously asking for knowledge related to the context of the user to new devices discovered. Also, if the user shows her interest in a specific location (e.g., downtown) or concept (e.g., hotel), OMs broadcast a message asking for knowledge related to it. An OM that receives a knowledge petition applies ontology modularization techniques (Stuckenschmidt et al. 2009) to extract relevant knowledge from its local ontology. When the new knowledge is received, OMs

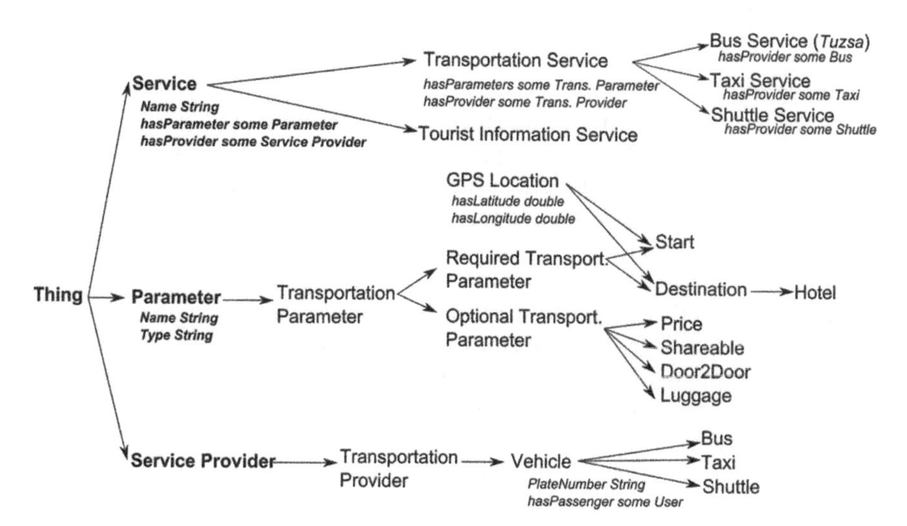

Fig. 4.13 Subset of an ontology that defines an LBS: "Transportation Service"

apply ontology matching techniques (Euzenat and Shvaiko 2007) to integrate the new knowledge into its local ontology. In this way, an OM will discover semantic relationships among terms such as synonymy (e.g., the terms "Taxi" and "Cab" are synonyms) or hypernymy (e.g., the term "Car" is a hypernym of the term "Taxi").

Generating an Information Request When the user shows her interest in a certain location, SHERLOCK on her device uses the knowledge in its local ontology to help her generate an information request. The goal of this step is to generate a request that SHERLOCK will later process to find the information that the user needs.

There are three (static) agents involved in this process: *ADUS, Alfred,* and *User Request Manager* (URM). ADUS is in charge of generating graphical interfaces appropriate to show information to the user and get her input. Alfred is in charge of storing information about the user and her device (e.g., user preferences, previous requests, and technical capabilities of the device). This information is especially relevant to infer interesting information from previous user requests when generating a new request. For example, Alfred can infer that the user usually selects taxis and shuttles when looking for transportation, and thus, she seems more interested in private transportation than in public transportation. Finally, URMs interact with the user to generate the final information request. Using the location that the user selected, the URM infers which LBSs are related to it. This is done by finding all the LBSs defined in the ontology that are related (e.g., through a property) to the type of location selected. For example, if the user selects a hotel, the URM will infer that services to book a room or to find a transport to go there can be interesting. The context of the user is also used to filter out some of these services (e.g., if the user is already in a taxi, the service to find transportation would probably not be interesting for her). Then, the URM presents the possible services to the user and when she selects one of them, the URM presents its parameters (if any) to find the most appropriate service provider for the selected service. Some of these parameters will be filled in by the URM using the information stored by Alfred.

Processing an Information Request Once SHERLOCK has generated an information request to find the most appropriate service providers for the user, the next step is processing it. Following the approach presented in LOQOMOTION (Ilarri et al. 2006b) to process location-dependent queries, SHERLOCK creates a network of mobile agents to find these service providers. These mobile agents, used as a way to balance the computing load and minimize the network latency, consider every device in the scenario as a potential processing node. Mobile agents continuously evaluate the appropriateness of the current device where they are executing for the task they are performing. As a result, these agents can stay in the same device, move to another one where the performance is expected to be better, or even create new helper mobile agents if they cannot solve the situation alone.

There are three types of mobile agents involved in the processing of an information request: *User Request Processor* (URP), *Trackers*, and *Updaters*. URPs are created to continuously process the user information request and return the results to their URMs (that will present these results to the user). URPs have to define the geographic area that would be interesting to obtain information related to the user request. Finally, a URP creates Tracker agents to monitor the selected area(s). A Tracker is in charge of monitoring an area to find relevant objects that can provide the service that the user needs. To achieve this goal, Trackers create a network of Updater agents and correlate their results. The Tracker is responsible for maintaining the network of Updaters trying to cover the area completely and with an adequate frequency (due to communication delays). Finally, Updaters are in charge of communicating with objects asking for the ontological context that describes them. Updaters are provided by their Trackers with the ontological description of the service providers interesting for the user. So, if an object's profile fulfills this description, then its information (especially its location) will be returned to the Tracker. Note that Updaters can use the reasoner on the device they are executing to check if an unknown object could be classified as an instance of the target provider it is looking for.

Prototype We have developed a prototype of the SHERLOCK (Yus et al. 2013b) as an Android app (see Figs. 4.14–4.16 for some screenshots of the prototype). The prototype uses the OWL API (Horridge and Bechhofer 2011), an ontology API to manage OWL 2 ontologies in Java applications, the JFact reasoner, and the SPRINGS mobile agent platform (Ilarri et al. 2006a). Using Semantic Web technologies (ontology APIs and DL reasoners) on current mobile devices are feasible, as we studied in Yus et al. (2013a) and Bobed et al. (2014). In addition, in Bobed et al. (2014), we evaluated the performance of semantic reasoners on smart devices and our results show that current smartphones can handle reasoning on small/medium ontologies.

First, the SHERLOCK prototype asks the user for some information such as her name and "profile" (e.g., person, researcher, taxi, or bus). Then, the prototype creates P2P networks using WiFi to communicate with other devices and to exchange OWL ontologies. Finally, the prototype helps users to create their information re-

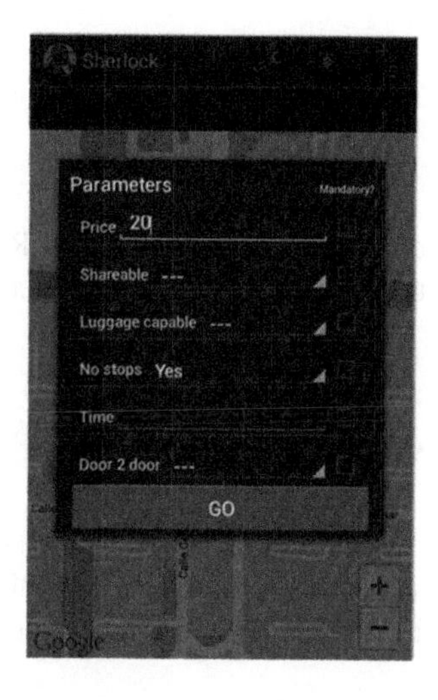

Fig. 4.14 The user fills in parameters to select the most appropriate service provider

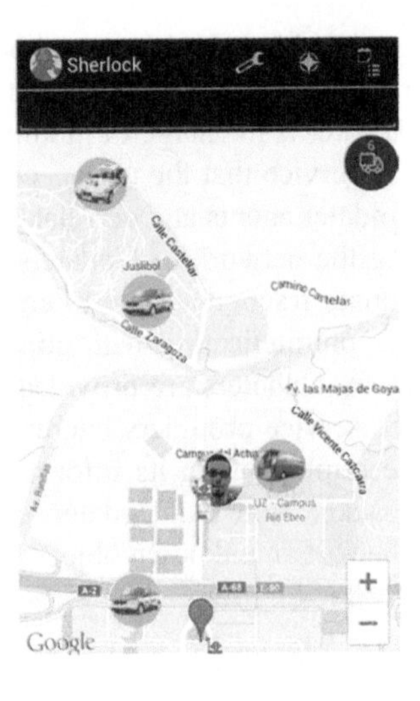

Fig. 4.15 Real-time location of service providers are shown on a map

Fig. 4.16 Different information requests can be processed for a user

quests and to find other SHERLOCK devices around whose "profile" matches with the kind of appropriate service providers for the user (Fig. 4.15).

4.6 Discussion

Throughout this chapter, we have presented several examples of how the use of semantic techniques leads to the development of smarter information systems. Making the computer aware of the semantics of the data will require still a long road of research in many fields. However, as we have shown, Semantic Web technologies in their current state make it possible to improve existing approaches, and to devise new and more intelligent applications.

In the SID research group, we are following this line of thought and research, attempting always to go a step further into exploiting semantics to improve the capabilities of our systems. We have seen how the use of semantic techniques has allowed us to improve Web searches, to perform semantic keyword-based searches on heterogeneous information systems, or to develop a platform to provide smarter LBSs, for example. However, our lines of research, which are directly implemented in our prototypes, are still open. Among others, some of the issues we are currently working on are:

- There is plenty of work to be done yet in the field of obtaining the semantics out from plain keywords and plain text. We are studying how to combine ontologies with NLP techniques to enhance this process, and solve many of the linguistic problems that exist (e.g., ambiguity and multilingualism). Moreover, with the know-how acquired during the development of QueryGen, Doctopush, and GENIE, we also want to study the open problem of Question Answering, which has recently attracted new attention (Lopez et al. 2011, 2013) due to the possibilities that ontologies provide to interpret the meaning of the posed queries.
- Regarding semantic searches and semantic LBSs, we want to study the inclusion of context information. User preferences could be introduced in our systems to provide better results in terms of more intelligent services providing more relevant results. For example, in some cases, some data could be provided to the user even in the absence of explicit queries, such as in *mobile recommendations* (Rodríguez-Hernández and Ilarri 2014). As another example, if we allow queries such as "retrieve hotels that are near," the notion of *near* is imprecise, and depends on the context (e.g., is the user walking or driving?) and the user preferences.
- In the context of location modeling, we also want to explore further the modeling capabilities of DLs to capture different spatial relationships and enhance our semantic location model. We will analyze how to introduce RCC relations other than *isContained* (e.g., inner and outer tangential relations), and how to model the dependency between all the RCC relationships. Our objective is to model RCC in a DL in such a way that, if the topology is possible, then the TBox with the axioms describing the topology is consistent and all topology relationships that we can derive from the TBox are realizable.
- Finally, we plan to study the semantic management of multimedia data in the context of SHERLOCK. Current mobile devices generate large amounts of photos and videos that SHERLOCK could take into account when offering the user with interesting information. We will explore the integration of our previous experience on multimedia information management (Yus et al. 2014b) with the semantic management of data of SHERLOCK.

The Semantic Web and its associated technologies are here to stay and are no longer restricted to web environments. Of course, there are still open issues and a long road to research in this field. For example, there is still no general purpose search engine that, given a query, returns the exact answer the user is looking for. For instance, a user that inputs the query "what is the meaning of life the universe and everything?" would expect the result to be just "42" and not thousands of web sites talking about The Hitchhiker's Guide to the Galaxy. Although some important knowledge bases have been generated in the last few years that might contain this information, we are still far from a scenario where computers understand the meanings behind any type of data and data repository. Nevertheless, in parallel with the enhancement of the current semantic techniques and Semantic Web technologies, we have shown that we can embrace and integrate them to change the way we devise applications.

Acknowledgment This work has been supported by the CICYT projects TIN2010-21387-C02-02 and TIN2013-46238-C4-4-R, and DGA-FSE.

References

Aho, A., Sethi, R., & Ullman, J. (2006). *Compilers: Principles, techniques, and tools*. New York: Addison-Wesley.

Baader, F., Calvanese, D., McGuinness, D., Nardi, D., & Patel-Scheneider, P. (2003). *The description logic handbook: Theory, implementation and applications*. New York: Cambridge University Press.

Banerjee, S., & Pedersen, T. (2003). *Extended gloss overlaps as a measure of semantic relatedness*. In Proceedings of the 18th International Joint Conference on Artificial Intelligence (IJCAI'03) (pp. 805–810). Acapulco, Mexico: Morgan Kaufmann.

Belussi, A., Combi, C., & Pozzani, G. (2009). *Formal and conceptual modeling of Spatio-temporal granularities*. In Proceedings of the 13th International Database Engineering & Applications Symposium (IDEAS'09) (pp. 275–283). Calabria, Italy: ACM.

Bernad, J., Bobed, C., Mena, E., & Ilarri, S. (2013). A formalization for semantic location granules. *International Journal of Geographical Information Science, 27*(6), 1090–1108.

Berners-Lee, T., Hendler, J., & Lassila, O. (2001). The semantic web. *Scientific American, 284*(5), 34–43.

Bird, S., Day, D., Garofolo, J. S., Henderson, J., Laprun, C., & Liberman, M. (2000). *ATLAS: A flexible and extensible architecture for linguistic annotation*. In Proceedings of the 2nd International Conference on Language Resources and Evaluation (LREC'00) (pp. 1699–1706).

Bizer, C., Heath, T., & Berners-Lee, T. (2009a). Linked Data—the story so far. *International Journal on Semantic Web and Information Systems, 5*(3), 1–22.

Bizer, C., Lehmann, J., Kobilarov, G., Auer, S., Becker, C., Cyganiak, R., & Hellmann, S. (2009b). DBpedia—a crystallization point for the web of data. *Web Semantics: Science, Services and Agents on the World Wide Web, 7*(3), 154–165.

Bobed, C. (2013). *Semantic keyword-based search on heterogeneous information systems*. PhD in Computer Science, University of Zaragoza, Spain.

Bobed, C., Ilarri, S., & Mena, E. (2010). *Exploiting the semantics of location granules in location-dependent queries*. In Proceedings of the 14th East-European Conference on Advances in Databases and Information Systems (ADBIS'10) (Vol. 6295, pp. 79–93). Berlin: Springer.

Bobed, C., Bobillo, F., Yus, R., Esteban, G., & Mena, E. (2014). *Android went semantic: Time for evaluation*. In Proceedings of the 3rd International Workshop on OWL Reasoner Evaluation (ORE'14) (pp. 23–29). CEURWS.

Bobillo, F. (2008). *Managing vagueness in ontologies*. PhD in Computer Science, University of Granada, Spain.

Bobillo, F., & Straccia, U. (2008). *fuzzyDL: An expressive fuzzy description logic reasoner*. In Proceedings of the 17th IEEE International Conference on Fuzzy Systems (FUZZ-IEEE 2008) (pp. 923–930). IEEE Computer Society.

Bobillo, F., & Straccia, U. (2011). Fuzzy ontology representation using OWL 2. *International Journal of Approximate Reasoning, 52*(7), 1073–1094.

Bobillo, F., & Straccia, U. (2012). Generalized fuzzy rough description logics. *Information Sciences, 189*(1), 43–62.

Bobillo, F., & Straccia, U. (2013). Aggregation operators for fuzzy ontologies. *Applied Soft Computing, 13*(9), 3816–3830.

Bobillo, F., Delgado, M., & Gómez-Romero, J. (2012a). DeLorean: A reasoner for fuzzy OWL 2. *Expert Systems with Applications, 39*(1), 258–272.

Bobillo, F., Delgado, M., Gómez-Romero, J., & Straccia, U. (2012b). Joining Gödel and Zadeh fuzzy logics in fuzzy description logics. *International Journal of Uncertainty, Fuzziness and Knowledge-Based Systems, 20*(4), 475–508.

Couclelis, H. (2010). Ontologies of geographic information. *International Journal of Geographical Information Science, 24*(12), 1785–1809.

Cunningham, H., Maynard, D., Bontcheva, K., & Tablan, V. (2002). *A framework and graphical development environment for robust NLP tools and applications*. In Proceedings of the 40th Annual Meeting of the Association for Computational Linguistics (ACL'02) (pp. 168–175). Philadelphia: ACL.

d'Aquin, M., Baldassarre, C., Gridinoc, L., Angeletou, S., Sabou, M., & Motta, E. (2007). *Characterizing knowledge on the semantic web with Watson*. In Proceedings of the 5th International Workshop on Evaluation of Ontologies and Ontology-Based Tools (EON'07) (Vol. 329, pp. 1–10). Busan: CEUR–WS.

Ding, H., Trajcevski, G., & Scheuermann, P. (2008). Efficient maintenance of continuous queries for trajectories. *Geoinformatica, 12*(3), 255–288.

Euzenat, J., & Shvaiko, P. (2007). *Ontology matching* (Vol. 18). Berlin: Springer.

Frank, A. (2003). *Chapter 2: Spatio-temporal databases*. Berlin: Springer Verlag.

Fu, H., & Anyanwu, K. (2011). *Effectively interpreting keyword queries on RDF databases with a rear view*. In Proceedings of the 10th International Semantic Web Conference (ISWC'11) (Vol. 7031, pp. 193–208). Springer.

Garrido, A. L., & Ilarri, S. (2014). *TMR: A semantic recommender system using topic maps on the items descriptions*. In The Semantic Web: ESWC 2014 Satellite Events (Vol. 8798, pp. 213–217). Springer.

Garrido, A. L., Gómez, O., Ilarri, S., & Mena, E. (2011). *NASS: News annotation semantic system*. In Proceedings of the 23rd IEEE International Conference on Tools With Artificial Intelligence (ICTAI'11) (pp. 904–905). IEEE Computer Society.

Garrido, A. L., Gómez, O., Ilarri, S., & Mena, E. (2012). *An experience developing a semantic annotation system in a media group*. In Proceedings of the 17th International Conference on Applications of Natural Language Processing to Information Systems (NLDB'12) (pp. 333–338). Springer.

Garrido, A. L., Granados-Buey, M., Ilarri, S., & Mena, E. (2013a). *GEO-NASS: A semantic tagging experience from geographical data on the media*. In Proceedings of the 17th East-European Conference on Advances in Databases and Information Systems (ADBIS'13) (pp. 56–69). Springer.

Garrido, A. L., Granados-Buey, M., Escudero, S., Ilarri, S., Mena, E., & Silveira, S. (2013b). *TM-Gen: A topic map generator from text documents*. In Proceedings of the 25th IEEE International Conference on Tools With Artificial Intelligence (ICTAI'13) (pp. 735–740). IEEE Computer Society.

Garrido, A. L., Peiro, A., & Ilarri, S. (2014a). *Hypatia: An expert system proposal for documentation departments*. In Proceedings of the 12th IEEE International Symposium on Intelligent Systems and Informatics (SISY'14) (pp. 315–320). IEEE Computer Society.

Garrido, A. L., Pera, M. S., & Ilarri, S. (2014b). *SOLER, a semantic and linguistic approach for book recommendations*. In Proceedings of the 14th IEEE International Conference on Advanced Learning Technologies (ICALT'14) (pp. 524–528). IEEE Computer Society.
Garrido, A. L., Granados-Buey, M., Escudero, S., Peiro, A., Ilarri, S., & Mena, E. (2014c). *The GENIE project—a semantic pipeline for automatic document categorisation*. In Proceedings of the 10th International Conference on Web Information Systems and Technologies (WEBIST'14) (pp. 161–171). SCITEPRESS.
Gedik, B., & Liu, L. (2006). MobiEyes: A distributed location monitoring service using moving location queries. *IEEE Transactions on Mobile Computing, 5*(10), 1384–1402.
Glimm, B., Horrocks, I., Motik, B., Stoilos, G., & Wang, Z. (2014). HermiT: An OWL 2 reasoner. *Journal of Automated Reasoning, 40*(2–3), 89–116.
Gómez-Pérez, A., Fernández-López, M., & Corcho, Ó. (2004). *Ontological engineering*. London: Springer.
Gracia, J., & Mena, E. (2008, Sept). *Web-based measure of semantic relatedness*. In Proceedings of the 9th International Conference on Web Information Systems Engineering (WISE'08) (Vol. 5175, pp. 136–150). Springer.
Gracia, J., & Mena, E. (2009). *Multiontology semantic disambiguation in unstructured web contexts*. In Proceedings of Workshop on Collective Knowledge Capturing And Representation (CKCaR'09).
Gracia, J., d'Aquin, M., & Mena, E. (2009). *Large scale integration of senses for the semantic web*. In Proceedings of the 18th International World Wide Web Conference (WWW'09) (pp. 611–620). ACM.
Granados-Buey, M., Garrido, A. L., Escudero, S., Trillo, R., Ilarri, S., & Mena, E. (2014a). *SQX-Lib: Developing a semantic query expansion system in a media group*. In Proceedings of the 36th European Conference on IR Research (ECIR'14) (Vol. 8416, pp. 780–783). Springer.
Granados-Buey, M., Garrido, A. L., & Ilarri, S. (2014b). *An approach for automatic query expansion based on NLP and semantics*. In Proceedings of the 25th International Conference on Database and Expert Systems Applications (DEXA'14) (pp. 349–356). Springer.
Gruber, T. R. (1993). A translation approach to portable ontology specifications. *Knowledge Acquisition, 5*(2), 199–220.
Gruber, T. R. (1995). Toward principles for the design of ontologies used for knowledge sharing. *International Journal of Human-Computer Studies, 43*(5–6), 907–928.
Haarslev, V., Lutz, C., & Möller, R. (1998). *Foundations of spatioterminological reasoning with description logics*. In Proceedings of the 6th international conference on principles of knowledge representation and reasoning (KR'98) (pp. 112–123). Morgan Kaufmann.
Harris, S., Seaborne, A., & Prud'hommeaux, E. (2013). SPARQL 1.1 Query Language. (http://www.w3.org/TR/sparql11-query). Accessed 20 April 2015.
Hightower, J. (2003). *From position to place*. In Proceedings of the 2003 workshop on location-aware computing (pp. 10–12). Springer.
Hitzler, P., Krötzsch, M., Parsia, B., Patel-Schneider, P. F., & Rudolph, S. (2012). OWL 2 web ontology language primer (Second Edition). (http://www.w3.org/TR/owl-primer). Accessed 20 April 2015.
Hoareau, C., & Satoh, I. (2007). *A model checking-based approach for location query processing in pervasive computing environments*. In Proceedings of the 2nd International Workshop on Pervasive Systems 2007(PerSys'07) (Vol. 4806, pp. 866–875). Springer.
Hoareau, C., & Satoh, I. (2009). *From model checking to data management in pervasive computing: A location-based query-processing framework*. In Proceedings of the ACM International Conference on Pervasive Services (ICPS'09) (pp. 41–48). ACM.
Horridge, M., & Bechhofer, S. (2011). The OWL API: A java API for OWL ontologies. *Semantic Web, 2*(1), 11–21.
Ilarri, S., Trillo, R., & Mena, E. (2006a). *SPRINGS: A scalable platform for highly mobile agents in distributed computing environments*. In Proceedings of the 4th International WoWMoM 2006 Workshop on Mobile Distributed Computing (MDC'06) (pp. 633–637). IEEE Computer Society.

Ilarri, S., Mena, E., & Illarramendi, A. (2006b). Location-dependent queries in mobile contexts: Distributed processing using mobile agents. *IEEE Transactions on Mobile Computing, 5*(8), 1029–1043.

Ilarri, S., Mena, E., & Bobed, C. (2007). *Processing location-dependent queries with location granules*. In Proceedings of the 2nd International Workshop on Pervasive Systems 2007(PerSys'07) (Vol. 4806, pp. 856–866). Springer.

Ilarri, S., Corral, A., Bobed, C., & Mena, E. (2009). *Probabilistic granule-based inside and nearest neighbor queries*. In Proceedings of the 13th East-European Conference on Advances in Databases And Information Systems (ADBIS'09) (Vol. 5739, pp. 103–117). Springer.

Ilarri, S., Mena, E., & Illarramendi, A. (2010). Location-dependent query processing: Where we are and where we are heading. *ACM Computing Surveys, 42*(3), 1–73.

Ilarri, S., Bobed, C., & Mena, E. (2011a). An approach to process continuous location-dependent queries on moving objects with support for location granules. *Journal of Systems and Software, 84*(8), 1327–1350.

Ilarri, S., Illarramendi, A., Mena, E., & Sheth, A. (2011b). Semantics in location-based services—guest editors' introduction for special issue. *IEEE Internet Computing, 15*(6), 10–14.

ISO/IEC. (2011). ISO/IEC 9075:2011 Standard, Information Technology—Database Languages—SQL.

Joachims, T. (1998). *Text categorization with support vector machines: Learning with many relevant features*. In Proceedings of the 10th European conference on machine learning (ECML 1998) (Vol. 1398, pp. 137–142). Springer.

Kaufmann, E., & Bernstein, A. (2010). Evaluating the usability of natural language query languages and interfaces to semantic web knowledge bases. *Web Semantics: Science, Services and Agents on the World Wide Web, 8*(4), 377–393.

Lange, D. B., & Oshima, M. (1999). Seven good reasons for mobile agents. *Communications of the ACM, 42*(3), 88–89.

Lopez, V., Uren, V. S., Sabou, M., & Motta, E. (2011). Is question answering fit for the semantic web? a survey. Semantic Web -Interoperability, Usability. *Applicability, 2*(2), 125–155.

Lopez, V., Unger, C., Cimiano, P., & Motta, E. (2013). Evaluating question answering over linked data. *Web Semantics: Science, Services and Agents on the World Wide Web, 21*(0), 3–13. (Special Issue on Evaluation of Semantic Technologies).

Lutz, M., & Klien, E. (2006). Ontology-based retrieval of geographic information. *International Journal of Geographical Information Science, 20*(3), 233–260.

Lutz, C., & Möller, R. (1997). *Defined topological relations in description logics*. In Proceedings of the 1997 international workshop on description logics (DL'97) (pp. 27–29). Morgan Kaufmann Publishers Inc.

Manning, C. D., Raghavan, P., & Schütze, H. (2008). *Introduction to information retrieval*. New York: Cambridge University Press.

Mena, E., & Illarramendi, A. (2001). *Ontology-based query processing for global information systems*. Boston: Kluwer.

Mendes, P., Jakob, M., & Bizer, C. (2012). *DBpedia: A multilingual cross-domain knowledge base*. In Proceedings of the 8th International Conference on Language Resources and Evaluation (LREC'12) (pp. 1813–1817). European language resources association (ELRA).

Miller, G. A. (1995). WordNet: A lexical database for English. *Communications of the ACM, 38*(11), 39–41.

Mokbel, M. F., Xiong, X., Hammad, M. A., & Aref, W. G. (2005). Continuous query processing of spatio-temporal data streams in PLACE. *Geoinformatica, 9*(4), 343–365.

Motik, B., Shearer, R., & Horrocks, I. (2009). Hypertableau reasoning for description logics. *Journal of Artificial Intelligence Research, 36*(1), 165–228.

Pepper, S., & Moore, G. (2001). XML Topic Maps (XTM) 1.0-TopicMaps.org Specification. (http://www.topicmaps.org/xtm). Accessed 20 April 2015.

Po, L. (2009). *Automatic lexical annotation: An effective technique for dynamic data integration*. PhD in Computer Science, Doctorate School of Information and Communication Technologies, University of Modena e Reggio Emilia, Italy.

Po, L., Sorrentino, S., Bergamaschi, S., & Beneventano, D. (2009). *Lexical knowledge extraction: An effective approach to schema and ontology matching*. In Proceedings of the 10th European Conference on Knowledge Management (ECKM'09) (pp. 617–626). Academic Publishing Limited.

Randell, D. A., Cui, Z., & Cohn, A. G. (1992). *A spatial logic based on regions and connection*. In Proceedings of the 3rd International Conference on Principles Of Knowledge Representation and Reasoning (KR'92) (pp. 165–176). Morgan Kaufmann.

Rigaux, P., Scholl, M., & Voisard, A. (2002). *Spatial databases with application to GIS*. San Francisco: Morgan Kaufmann.

Rodríguez-Hernández, M. C., & Ilarri, S. (2014). *Towards a context-aware mobile recommendation architecture*. In Proceedings of the 11th International Conference on Mobile Web Information Systems (MobiWIS'14) (pp. 56–70). Springer.

Russell, S., & Norvig, P. (2003). *Artificial intelligence: A modern approach*. Englewood Cliffs: Prentice-Hall.

Schiller, J., & Voisard, A. (2004). *Location-based services*. San Francisco: Morgan Kaufmann.

Sekine, S., & Ranchod, E. (Eds.). (2009). *Named entities: Recognition, classification and use*. Amsterdam: John Benjamins.

Shadbolt, N., Hall, W., & Berners-Lee, T. (2006). The semantic web revisited. *IEEE Intelligent Systems, 21*(3), 96–101.

Shekhar, S., & Chawla, S. (2002). *Spatial databases: A tour*. Upper Saddle River: Prentice Hall.

Sirin, E., Parsia, B., Grau, B. C., Kalyanpur, A., & Katz, Y. (2007). Pellet: A practical OWL-DL reasoner. *Web Semantics: Science, Services and Agents on the World Wide Web, 5*(2), 51–53.

Sistla, A. P., Wolfson, O., Chamberlain, S., & Dao, S. (1997). *Modeling and querying moving objects*. In Proceedings of the 13th International Conference on Data Engineering (ICDE'97) (pp. 422–432). IEEE Computer Society.

Smeaton, A. F. (1999). Using NLP or NLP resources for information retrieval tasks. In T. Strzalkowski (Ed.), *Natural language information retrieval* (Vol. 7, pp. 99–111). Dordrecht: Springer.

Spyrou, C., Samaras, G., Pitoura, E., & Evripidou, P. (2004). Mobile agents for wireless computing: The convergence of wireless computational models with mobile-agent technologies. *Mobile Networks and Applications, 9*(5), 517–528.

Stuckenschmidt, H., Parent, C., & Spaccapietra, S. (2009). *Modular ontologies: Concepts, theories and techniques for knowledge modularization* (Vol. 5445). New York: Springer.

Trillo, R., Gracia, J., Espinoza, M., & Mena, E. (2007a). Discovering the semantics of user keywords. *Journal on Universal Computer Science, 13*(12), 1908–1935.

Trillo, R., Ilarri, S., & Mena, E. (2007b). *Comparison and performance evaluation of mobile agent platforms*. In Proceedings of the 3rd International Conference on Autonomic and Autonomous Systems (ICAS'07) (pp. 41–46). IEEE Computer Society.

Trillo, R., Po, L., Ilarri, S., Bergamaschi, S., & Mena, E. (2011). Using semantic techniques to access web data. *Information Systems, 36*(2), 117–133. (Special Issue: Semantic Integration of Data, Multimedia, and Services).

Wache, H., Voegele, T., Visser, U., Stuckenschmidt, H., Schuster, G., Neumann, H., & Hübner, S. (2001). *Ontology-based integration of information—a survey of existing approaches*. In Proceedings of the IJCAI Workshop: Ontologies and Information Sharing (pp. 108–117). CEUR–WS.

Yus, R., Bobed, C., Esteban, G., Bobillo, F., & Mena, E. (2013a). *Android goes Semantic: DL reasoners on smartphones*. In Proceedings of the 2nd International Workshop on OWL Reasoner Evaluation (ORE'13) (pp. 46–52). CEUR–WS.

Yus, R., Mena, E., Ilarri, S., & Illarramendi, A. (2013b). *SHERLOCK: A system for location-based services in wireless environments using semantics*. In Proceedings of the 22nd International World Wide Web Conference (WWW'13) (pp. 301–304). International World Wide Web Conferences Steering Committee.

Yus, R., Mena, E., Ilarri, S., & Illarramendi, A. (2014a). SHERLOCK: Semantic management of location-based services in wireless environments. *Pervasive and Mobile Computing, 15*, 87–99.

Yus, R., Mena, E., Ilarri, S., Illarramendi, A., & Bernad, J. (2014b). *Multi-CAMBA: A system for selecting camera views in live broadcasting of sport events using a dynamic 3D model*. Multimedia Tools and Applications, 32 pages. Published online: 15 December 2013.

Zadeh, L. A. (1965). Fuzzy sets. *Information and Control, 8*(3), 338–353.

Chapter 5
Semantics: Revolutionary Breakthrough or Just Another Way of Doing Things?

Andrew W. Crapo and Steven Gustafson

5.1 Introduction

Today, we see several new trends regarding data management and use. First, data is being produced by many more people and things. Governments and consumers have entered the business of producing data alongside industry. The data.gov initiative is a perfect example of how governments are enabling and promoting data sharing and reuse in many interesting ways. The data.gov initiative was created in 2009 by the US government with the goal of making data more available to the public in a format that both people and machines could read: XML (data.gov 2014). Activities from the consumer side can be seen in initiatives around the quantified self and web of things (Gustafson and Sheth 2014). Data looks different than it used to: from tweets to health-care records, and from activity monitors to multimodal and multidimensional flight data, we see data coming in new and different formats that need to be analyzed alongside each other. However, the most often talked about attribute of today's data is that it is produced at volumes of increasing size, i.e., the era of big data.

Second, the users of data are becoming more diverse and complex. The most recognizable trend here is the democratization of data analysis via the new tools, new training and curriculum, and the recognized role of the data scientist within organizations. Third, data is being stored in newer and more diverse infrastructures due to the confluence of increased volume and diverse users. Cloud services, like Amazon Web Services, are proliferating to bring down cost while driving up flexibility and reliability. Fourth, and finally, we are entering a new era of applications that feed off data to create even more valuable data, namely artificial intelligence and the automation of a variety of human tasks. In these areas, like driverless cars, superhuman robots (DARPA Grand Challenges), and intelligent personal assistants

A. W. Crapo (✉) · S. Gustafson
GE Global Research, K1 3A26, 1 Research Circle, Niskayuna, NY 12309, USA
e-mail: crapo@research.ge.com

M. Workman (ed.), *Semantic Web*, DOI 10.1007/978-3-319-16658-2_5

(e.g., Siri) and information retrieval (e.g., IBM's Watson), data provides not only the critical knowledge bases on which to reason and act but also the surrogate to perception via video, radar, and other sensors. Given these four very real trends, we believe that the Semantic Web, or more recently named Linked Data (Bizer et al. 2009), developed and promoted over the past decade and a half, is poised to become a critical, enabling technology from knowledge capture to knowledge generation, and from data interoperability to data integration.

Recently, we were invited to a meeting in which a team of relational database design experts were being familiarized with a "data problem." Real-time data flowed from people and machines in a manufacturing facility into data repositories. The problem was how to enable quality engineers to retrieve the right data on which to perform analytics in an effort to identify possible causes of manufacturing defects. We described to the group the basic ideas of semantic technology as envisioned by the Semantic Web and Linked Open Data. The lead expert listened politely and finally said, with some exasperation, "How is this different? I can do all of the things you're describing with relational technology." It was not the first time that we felt that we had somehow failed to make the case. Was there a real difference? Or was semantic technology just another way of doing the same thing?

This experience, alongside the aforementioned trends and others like them, has provided the motivation for selecting the topic for this chapter. Our objective is to identify the foundation and capabilities of semantic technologies in a broad sense, as well as the more specific sense of Semantic Web technologies, and compare and contrast them with relational database technology, with XML, with object-oriented programming languages, with Unified Modeling Language (UML), and with the broader spectrum of logic and logic programming. This chapter is not an unbiased weighing of all the technologies. Rather it tries to answer the question of the chapter title while identifying strengths and weaknesses of these modeling approaches. And we are evangelists of semantic technology, so we will be trying to make the case that semantic technology is better in important ways. The chapter ends with examples of how semantic technology has helped us in our work to further strengthen our argument. This argument is predicated on the notion that we need to return to "meaning" in our data management and use technologies, that over time we have moved away from the "meaning," and that by making this shift we will be in the best position to address the trends of increased data production, use, and infrastructure. By capturing the "meaning" of data in flexible models that enable rich inference and implication, we will allow minds and machines to grasp both the meaning and the implications of the data. In other words, we need to merge data management and knowledge management into a synergistic whole.

The Data World Today: Relational Plus

A phrase that was used to promote big data technology is particularly insightful when considering the state of the art of data management and use: *Big data technologies put the data first*. What is being communicated by the phrase is the idea that data management and infrastructure do not necessarily put the data "first," or as the most important thing, in today's applications and use of data. We believe that the evolution of the modern IT department and its separation from the business applications and data users evolved from the optimization of two different criteria. The latter, the business applications, desire to create the most insightful and value-driving use of data. However, the former, the IT departments, optimize for lowering the cost of ownership and support for data use. Consequently, standard technologies in IT are usually adopted, and there is a widening divide between the users of data and the teams tasked with maintaining the data. Big data stressed both the cost of ownership of data by challenging the disk space, compute and networking, as well as the ability to leverage data in a value-adding way since big data allows more people to use the data in more complex ways. In essence, big data technologies attempted to solve the cost of ownership problem by enabling inexpensive commodity hardware to be cheaply and robustly networked, and by forming a very simple model of the data (key–value pairs) that most data-savvy users and applications could easily understand. We argue that while big data technologies succeeded in their first objective, the second objective of a simple data model (key–value pairs) was short of today's needs.

Relational database technology has been a foundation of information systems for decades. Its conceptual roots go back to the work of Edgar F. Codd in 1969. It addressed the critical need to store data in a manner which would allow use by multiple applications, including applications not yet envisioned. It was well founded in relational theory, and this allowed for the development of powerful query (e.g., Structured Query Language, SQL) and manipulation (e.g, Procedural Language, PL/SQL) languages. It also matched the computational capability, given the size of datasets available at the time. The result was a rich array of software and hardware that provides rapid and reliable data management.

Over time the quantity and complexity of data expanded significantly and began to push the limits of relational technology. Solutions included creating data marts—specialized data warehouses for reporting or other purposes—and creating temporary "views" of the data. Both approaches essentially provide a pre-computation and retrieval of results to reduce computational demands at query time for better response times. However, as data sizes increased, and the need to provide certain database functionality like ACID compliance decreased in certain applications, traditional relational databases began to look over-expensive, over-featurized, and too slow[1]. Google paved the way for the future of Hadoop with its Big Table and Map Reduce. Apache took up the open source Hadoop project from Yahoo, which led to a whole new ecosystem of big data technologies. However, as the demand for and use of these low-cost technologies based on commodity systems grew, organizations and applications began to require more traditional database features to leverage existing database skillsets and provide guarantees around performance

[1] ACID is an acronym for atomicity, consistency, isolation, and durability, a set of properties that guarantee that database transactions are processed reliably. Some of the challengers to relational technology have emerged from applications such as social media where performance and scalability have at least initially seemed more important than transactional integrity.

and transactional consistencies, and to expect data models that allow applications to more easily tap into and integrate data in big data infrastructures. Big data technologies put "data first" by not requiring a lot of up-front data models and schemas, but as applications on big data technologies matured, the need to find "meaning" in all the data became more evident. For the first time in decades, the supremacy of relational is being challenged on multiple fronts.

What Would be Better? And by what Criteria?

As noted by David Marr, every representation brings some aspects of the information to the forefront, making it more easily perceived, at the cost of pushing other information into the background, making it less easily perceived (Marr 1982). It follows that the objective selection of an information/knowledge representation will depend upon what information one desires to make more easily perceived. Arguably, this applies rather specifically to visual representations, but one cannot represent visually what is not present in the underlying data/knowledge. In the work that we do, some of which is described briefly at the end of this chapter, our priority is to make the domain-related aspects of the data easily perceivable and understandable by a subject matter expert (SME) or someone with reasonable domain knowledge, and to make the data "speak for itself" in the language of the domain. And we aspire to have the data speak to machines as well as minds[2].

Evaluation Criteria To that end, the following criteria are adopted for technology evaluation in this chapter. These criteria are not independent but each emphasizes a different aspect of what we consider to be important attributes of an information/knowledge representation and storage capability. We focus on the representation much more than the storage, believing that many storage schemes can be envisioned for a given representation, and that the representation is more crucial for expressive power and human/machine understanding while the storage can be tailored to the scale and performance requirements of the application.

1. How well does the representation capture a domain's data and knowledge in a form compatible with the mental models of, and therefore more easily understood by, SMEs and those with basic domain knowledge?
2. How easily can information be retrieved by the same users?
3. How expressive is the representation for enabling computation of new information from existing information; that is, enabling reasoning and inference?
4. How well does the representation support interoperability between disparate information systems?
5. Can data expressed in the representation "speak" to minds and machines with a meaning shared by both?

[2] *Minds and Machines* is the name of a journal associated with the Society for Machines and Mentality. It is the name of an MIT OpenCourseware offering. It is also the name of an annual GE-sponsored event (see https://www.gemindsandmachines.com/).

We use the above criteria as a framework to assess how well various representations/technologies are suited to the needs of today's data world. Those needs include that of making the meaning of data clear, which enables greater value to be derived from the data. We demonstrate, both theoretically and practically, that semantic technology is, as the name suggests, better at capturing data in ways that preserve, and even enable addition derivation of, meaning.

What We Mean by Semantic Technology?

First, let us reduce ambiguity by explaining what we mean by semantic technology. In the broadest sense, semantic technology is any representational paradigm that enables us to capture data and information in a way that enables computers as well as people to "understand" the meaning and not just work at a syntactic level. For example, XML is a technology that enables data to be represented in a standardized, domain-independent syntax. If I use my computer to express my data in standards-compliant XML, your computer will be able to parse the data. XML achieves syntactic interoperability.

But what does the data mean? And how would we enable a machine to "understand" the data? If we could provide answers to these questions, we would have achieved semantic interoperability. Consider the snippets of XML shown in Table 5.1.

Asked what the data in the first column of Table 5.1 is about, you might respond that it is the name and email address of one of the chapter authors. How did you arrive at that meaning? While it is not our intent to go deeply into semiotics, we will argue that the meaning you gave for the data was derived almost entirely from the fact that the tags used in the example, as well as the data itself, are words or symbols with which you are familiar and which coincide with concepts in your mental models. To illustrate, consider a second XML snippet with an identical structure shown in the second column of Table 5.1.

Note how different is your constructed interpretation. The meaning is not in the structure, as both snippets have the same structure. The tags (*name, cart,…*) and symbols (*@, $*) do not directly provide any meaning, but rather meaning comes from information about the tags and symbols that the reader supplies but is entirely missing from the representation. Nor does XML schema lend much assistance in

Table 5.1 Two XML snippets with parallel structure

<xml...>	<xml...>
....	....
<name>	<cart>
<given-name>Andrew</given-name>	<item>Tom Sawyer</item>
<surname>Crapo</surname>	<price>$19.95</price>
</name>	</cart>
<email> crapo@research.ge.com </email>	<total>$25.98</total>
....	....
</xml>	</xml>

capturing the meaning of the structure. XML schema provides a specification of the required syntax. In the Table 5.1 examples, the XML schema might indicate that the cart can contain many item/price pairs while it might prohibit that a name "contain" more than one given name/surname pair. Such syntactic constraints may devolve from the domain but they are not, of themselves, sufficient to enable interpretation of the meaning of the data.

So what do you have in your head that allows you to give meaning to the *<name>*, *<cart>*, and other tags and symbols and thereby give an interpretation to the data structures in these examples? And could we construct something that would do the same thing for a machine? One theory is that as we learn a language we create, in our minds, a directed graph relating different concepts, with associated words and/or phrases, in a variety of ways. The construction of this conceptual map is observable in young children as they add words to their vocabulary and gradually refine their meaning and usage—usage being about language syntax as well as subtle shades of meaning (Pan and Gleason 1997).

This process can be likened to the structuring of the tags used in an XML document. In the first example, such structuring might include relationships such as: (a) given name and surname are subclasses of name; (b) in certain cultures, a name often consists of at least two parts, a given name and a surname; and (c) name and email (address) are pieces of information about a person; hence, it might be inferred that *<person>* is the parent tag of both although such a tag is not shown in the snippet. These relationships can be captured in a computational model called an ontology. In fact, one definition of ontology is structured metadata. For XML, this might be the defining of the tags and the relating of the tags to one another. Meaning, it can be argued, is an emergent property of such structuring. Capturing this structuring in a computational model would in a significant sense allow a computer to "understand" what the data means.

There are numerous ways to structure metadata and thereby capture meaning. How flexibly and expressively a particular language or representation captures such structuring is a key determinant in how good it is as a semantic language or representation. The capturing of semantics is very closely akin to knowledge capture, and semantic technologies draw from predecessors in the knowledge representation domain. Most semantic representations are, at a minimum, fragments of a first-order predicate logic. This is discussed in greater detail in a subsequent section of the chapter. Predicates with arity 2, meaning that they can take two arguments, e.g., child(Hillary, Chelsea), can be seen as one segment of a directed graph, where a segment consists of a node, a directed edge, and a node, the three together being called a triple and consisting of a subject, a property or predicate, and an object. For example, the previous predicate and arguments are equivalent to the triple Hillary → child → Chelsea. Binary predicates can be domain dependent, e.g., "child," but they can also indicate domain-independent types of relationships between domain concepts or even a relationship to a concept of the language itself: child → type → property, Hillary → type → Woman, Chelsea → type → Woman, Woman → subclass-of → Person, Person → type → class. We refer to a language with predicate arity limited to no more than two as a graph-based language because any set of statements in the language constitutes a mathematical directed graph (and can be perceived as a graph as well).

Foundational Concepts in Modeling

In the preceding paragraphs, we have made passing reference to several ideas that need to be further clarified in order to establish a foundation for technology evaluation and comparison. These range from the question of how we synthesize meaning in our minds to rather casual references to "model" and the "relationships" between "concepts." In an effort to give more depth to a shared understanding of these and related ideas, we take a small tour into the realm of modeling.

We will not try to give an entirely formal definition to the term "model" because there are so many kinds of models. By model, we mean something which is similar, presumably in some useful or interesting way, to something else. Usually, a model is simpler than the thing being modeled, and it derives its usefulness, in part, from this simplification. Often a model represents something in the real world, and allows us to explain what happens or to predict what is expected to happen in that world. The "real world" can be stretched to include constructs such as financial markets or social media, expanding the domains in which we build and use models.

One may suppose that the first models constructed by human kind were those constructed in the mind—mental models. There is not a universally accepted theory of mental models, but one theory that has demonstrated explanatory and predictive merit, and which is useful for our present purposes, is the mental model theory developed by Johnson-Laird and his students at Princeton (Johnson-Laird 1983, 1988, 1993). What is intuitively attractive about this theory is its apparent consistency with the observable learning pattern of children, and the approach that people seem to naturally take to making sense of their observations.

Presented with a barrage of sensory input, we appear to identify patterns in that input and store the patterns in preference to storing all of the input. That certainly makes sense when thought of from a data management perspective[3]. This pattern identification, sometimes called abstraction or generalization, can conveniently be thought of as being of two types. The first is the grouping of similar "things" into named concepts, e.g., *plant, animal, tree, rock, person*, etc. This kind of grouping finds formalization in set theory. The second kind of pattern identification is in the relationships perceived between things. We give these relationship patterns names as well: the book is *on* the shelf, the man *caught* the fish, the woman *loves* the child. The second kind of grouping finds formalization in the predicates of predicate logic with arity greater than or equal to 2. Such a predicate captures a relationship between two or more things. However, both grouping patterns and

[3] But beware, we are not the first scientific epoch to view the working of the human mind as an extension of current technology. In a time when fluid dynamics was the newest thing, there were elaborate theories of how the brain worked based on fluid flows (Wade and Swanston 1991).

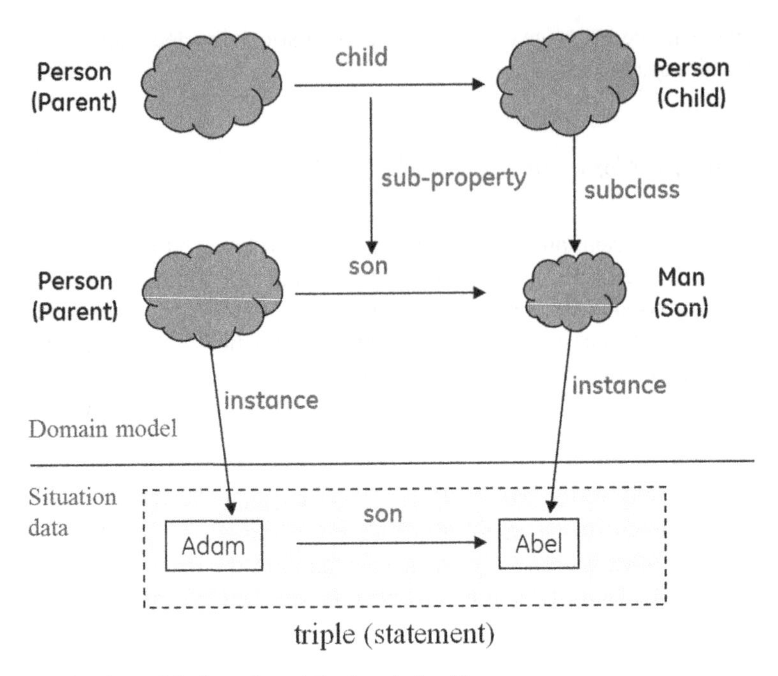

Fig. 5.1 A simple model of people and family relationships

relationship patterns are constructs of the mind—they do not exist, per se, in the "real world."

Another way of describing our use of the term "model" is to say that a model consists of a specific set of patterns of these two types relevant to some domain of interest. We extend our conceptualization of a model to include actual occurrences of these patterns: Data and observations consisting of either individual instances belonging to a group or multiple instances recognized to be in a relationship. We strongly prefer that an "instance data model" always include or make unambiguous reference to the definition of the relevant patterns or abstractions, and identify to which patterns any particular occurrence conforms. For example, if a model has the information, "John is a man who owns a black dog," it should include or have an unambiguous reference to where we find the meaning of "man," "owns," "black," and "dog" as well as the associations between the data and the meaning, e.g., between "John" and "man." Such associations make the data self-describing. Thought of another way, the context of the data is preserved with the data by maintaining associations between the data and the meta-model.

To make this a little less abstract, consider the simple model partially illustrated in Fig. 5.1. The starting point is the specification of the class *Person* and the property *child* with domain *Person* and range *Person.* Introducing the concept of the property *gender* with domain *Person* and range *Gender*, a class defined extensionally as either *Male* or *Female*, one can now define two subclasses of *Person, Man* and *Woman*. These class definitions are intensional (distinguished by their properties) and given by the following axioms:

> A *Person* is a *Man* if and only if *gender* is *Male*.
> A *Person* is a *Woman* if and only if *gender* is *Female*.

We might further clarify these classes by saying that *Man* and *Woman* are disjoint. We can then define the *son* and *daughter* properties as sub-properties of *child*, with ranges *Man* and *Woman*, respectively. This means that the values of the properties *son* and *daughter* are each subsets of the values of the *child* property, which is consistent with the definitions of *Man* and *Woman* as subclasses of *Person*.

Additional concepts can be added to the model. For example, *parent* is the inverse property of *child*, meaning:

> for all x and for all y, if *child* (x, y), then *parent* (y, x).

A useful extension of this concept is that of the transitive property *ancestor* with partial definition:

> for all x, for all y, and for all z, if *parent* (x, y) and *parent* (y, z), then *ancestor* (x, z).

To say that a property p is transitive means that:

> for all x, for all y, and for all z, if $p(x, y)$ and $p(y, z)$, then $p(x, z)$.

Our model might include role classes, meaning that the class is defined as the set of individuals that are values of a particular property. For example, the role class *Child* contains all individuals which are objects of the *child* property. We might also define *Parent* with a cardinality restriction on the *child* property:

> A *Person* is a *Parent* if and only if *child* has at least one value.

Similar role definitions could be given for the classes *Son, Daughter, Ancestor*, etc. Role classes are shown in parenthesis in Fig. 5.1. The distinction between a role class and an intensional class definition can be illustrated by two meanings that we might give to a class *Child*. We might mean the role class, i.e., any individual (of type *Person*) who is the value of a *child* property. However, we could also define *Child* intensionally as a *Person* with *age* less than 14. Natural language does not distinguish these two meanings with different words but leaves the meaning to be gleaned from context.

We could go on to give logical axiomatic definitions of properties such as *aunt, uncle, cousin, first cousin, first cousin once removed*, etc., and classes *Minor* (a *Person* with *age* <18), *Aunt, Uncle, Cousin*, etc. We might define *father* and *mother* properties and say that they are inverse functional, meaning that a *Person* can have only one *Father* and one *Mother*. We could define the property *sibling* and say that it is symmetric. The result would be a meta-model with increasing complexity of structure and associated richness of meaning. Given a dataset that identified some class members, e.g., *Man(Adam), Man(Abel)*, etc., and relationships, e.g., *child(Adam, Abel)*, etc., the computer would be able to infer many other class memberships and relationships, answer questions, and detect inconsistencies. We argue that such a model would allow a machine to have an understanding of data that is shared with human minds.

Importance of Similarity

While a model's simplicity over the thing modeled can be important, the type and degree of similarity between the model and that which is modeled is essential to the model's usefulness. The identification of sets and relationships as the key abstractions of a model provide us with a means of measuring this degree of similarity. When every set (class or type) in the model has one and only one corresponding set in the modeled world, and every relationship in the model has one and only one corresponding relationship in the modeled world, the mapping between model and modeled is isomorphic. Such a model has perfect similarity. That such perfect similarity is not usually achieved is reflected in the adage, "All models are wrong; some are useful." While work has been done to formalize the quantification of similarity (Gurr 1997), we use the term with less formality in this chapter. Figure 5.2 illustrates the comparison of the modeled world and the model in terms of their similarity.

Importance of Model Theory

Besides simplicity and similarity, a third highly desirable characteristic of a model is shown in Fig. 5.2 —the ability to determine whether any given statement in the model is true or false. If the modeling language has a model theory, it is said to have model-theoretic semantics. A model with a sound model theory is sometimes referred to as a formal model. For such a model, the theory, along with the associ-

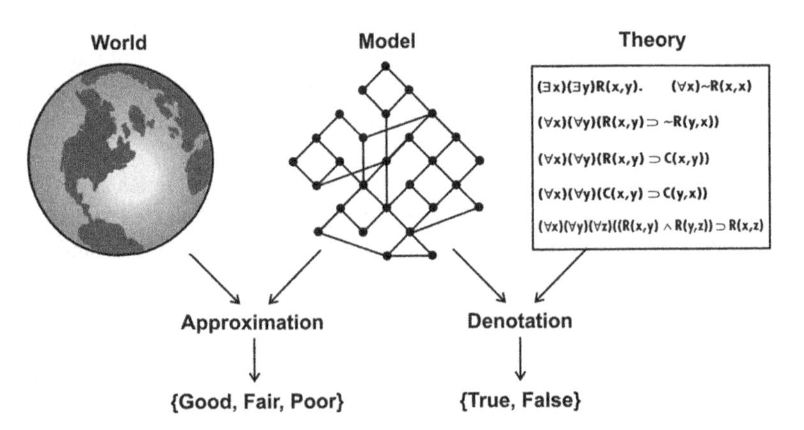

Fig. 5.2 The world, a model, and a theory. (Sowa 2014)

ated proof theory, provides a mechanizable means of not only evaluating the truth of any statement in the model but also doing truth-preserving transformations that produce new model content—content that was implicit in the model but can be made explicit via inference, also known as logical entailment. As noted above, set theory and predicate logic formalize what people naturally and informally do when they identify patterns in their observations and draw inferences from those patterns. Set theory and logic provide a means of evaluating statement truth in a model representation that uses them as a foundation.

For models lacking a model-theoretic semantics, the evaluation of the truth of model statements (content) and the accomplishment of truth-preserving transformations become a question of writing model-specific code. Such code abounds in the form of C++, Java, PL/SQL, etc. Most of this code is procedural, meaning that it is a set of instructions to be followed by the machine, but of the code's "meaning" the machine has no "understanding."

Logic and Logic Programming as Reference

Before evaluating various representations, let us briefly review logic and logic programming as a reference point. Logic, the science of valid inference, has fascinated mankind since ancient Greece, China, and India. A summary of logic can be done more neatly than the development actually occurred. Propositional logic deals with the truth of statements given the truth of other statements. It develops critical machinery for deriving new statements, or inferences, from existing statements asserted or previously inferred to be true or false. In propositional logic, a statement is atomic—there is not decomposition of a statement into its parts nor is there a way to determine the truth or falsehood of a statement from any internal considerations. Propositional logic consists of just four basic operations: conjunction (or), disjunction (and), negation (not), and implication (implies).

Predicate logic adds the essential capability to express the internal structure of a statement, and in the process a mechanism for evaluating the truth of the statement from its internals. It adds predicates, functions, and quantification—universal (for all…) and existential (there exists some…). A quantifier expresses that some variable ranges over some set of things, e.g., "for all x where x is a Person" means that whatever follows is true for every member of the Person class. In first-order logic, quantifiers are not allowed to range over predicates, which is equivalent to saying that sets of sets are not allowed. Second-order and higher-order logics relax this constraint.

In addition to higher-order logics, extensive work has been done in modal logic, which qualifies statements with modal operators such as belief, possibility, or temporal state. Each of these logics has a model and proof theory, although these become increasingly complex as expressiveness is added. Again, added complexity means more complex model and proof theories.

Even first-order logic is not decidable, meaning that applying the proof theory to a statement may not determine if it is true or false in finite time. For the past several decades, logicians have been hard at work identifying a system for classifying various flavors of logic and determining the computational characteristics of each (Baader et al. 2003). This work is leveraged by the Web Ontology Language (OWL) with different dialects implementing various fragments of first-order logic with differing computational requirements (Motik et al. 2012).

A number of languages implement logic in various ways. Prolog is a primarily declarative logic programming language rooted in first-order logic. Common logic (CL) is a relatively recent standard which shows great promise and unifies previous work in the Knowledge Interchange Format (KIF) and conceptual graphs (Information technology—Common Logic Standard (CL): a framework for a family of logic-based languages 2007). CycL is a language with elements from higher-order logic that has been developed by Cycorp since 1984. Its implementers have tried to create a system capable of showing some level of "common sense" and able to learn from reading natural language texts (Cyc 2014). This has remained an elusive goal.

We will not try to evaluate the various families of logic against our criteria. Arguably, each criterion might find an optimized logic, and there would surely be one that generally met the criterion better than the current generation of Semantic Web languages. However, wide adoption and the availability of affordable tools and open systems are of inestimable value, hence the focus of this book. The general capabilities of logic languages will be useful to us as a reference point in the following sections.

RDF, RDFS, and OWL as Semantic Languages

While many languages would qualify as semantic languages, by the description of semantics above, those used in the Semantic Web deserve special consideration given the topic of this book. Let us briefly review Resource Description Framework (RDF), RDF schema (RDFS), and OWL and relate them to the topics we have so far discussed.

RDF "is a standard model for data interchange on the Web" (Resource Description Framework (RDF) 2014). An RDF dataset consists of a default graph, which is unnamed and may be empty, and zero or more named graphs. Each graph is made up of a set of statements, each consisting of a subject, a predicate, and an object and called a triple. The elements of a triple can be an internationalized resource identifier (IRI), a blank (unnamed) node, or a literal. Literals may be typed. Each of these elements bring into focus important aspects of Semantic Web-style modeling.

- IRI—IRI is a generalization of uniform resource identifier (URI). As the last word suggests, these are used for identity in the Semantic Web. An IRI can be thought of as consisting of two parts, a namespace and a local name or fragment, sometimes separated from the namespace by the "#" symbol. For example, "http://www.w3.org/1999/02/22-rdf-syntax-ns#type" is the URI of the RDF property used to relate a resource to a set (class, group) to which it belongs. The local fragment is "type" and the namespace is "http://www.w3.org/1999/02/22-rdf-syntax-ns." To make RDF data more readable by people, the namespace is sometimes replaced with an associated prefix, which is then separated from the local fragment by a colon, e.g., "rdf:type." The predicate of an RDF statement is an IRI.

- Blank node—many (perhaps most) things in the word do not have unique identifiers, and to create such for them can be quite laborious and subject to pitfalls. These unnamed things are naturally referenced by their relationship with other things, e.g., "John's pen." Even those things which do have identifiers are often referenced by their relationships. "Let's take Sally's car" as opposed to "Let's take VIN-WPOZZZ99ZTS392124." RDF allows an unidentified resource to remain unidentified except by its relationships. The subject of an RDF statement can be an IRI or a blank node.
- Typed literal—many programming languages differentiate between those things which have identity, explicitly or by relationship, and "data values" for which one might argue that identity does not make sense. For example, if I have 12 eggs and your son is age 12, is that the same 12 or a different 12? XML schema defines a number of data types: string, boolean, decimal, date, time, duration, etc. In addition, applications can create user-defined data types. A typed literal is a value and its data type. The data type includes the value space, which specifies what values are valid for that type. The object of an RDF statement can be an IRI, a blank node, or a literal.

RDF by itself is a relatively weak modeling language. RDFS extends RDF in important ways, and it provides the basic elements for constructing an ontology—a model of what exists or can exist in a domain. RDFS adds the concept of a class (*rdfs:Class*), thereby explicitly adding the foundational concept of sets. The RDFS properties *subClassOf* and *subPropertyOf* allow classes and properties to be partially ordered into hierarchies, thereby enabling the concepts of subsets and super sets from set theory to be applied. Properties can be given a domain and range, thereby specifying the possible subjects and objects of an RDF statement using that property as predicate. Taken together, RDF and RDFS vocabularies allow a model-theoretic semantics for RDF graphs. This is the "theory" part of Fig. 5.2. It allows truth preserving transformations, also known as entailments or inferences, to occur, resulting in the derivation of new RDF statements from existing RDF statements.

OWL is a family of knowledge representation or ontology languages. OWL further extends the expressiveness of RDF and RDFS. OWL Full is the most expressive OWL dialect, and is not, in general, decidable with respect to truth-preserving transformations. Other flavors of OWL are decidable fragments of first-order logic based on description logics (DL). These flavors have different semantics and therefore support different truth-preserving transformations. In general, the additional expressiveness of OWL includes the following:

1. Universal quantification (see the previous section on logic) expressed by the *owl:allValuesFrom* class axiom
2. Existential quantification (see the previous section on logic) expressed by the *owl:someValuesFrom* class axiom
3. The restriction of a property of a class to a particular value with the *owl:hasValue* class axiom
4. Cardinality restrictions on a property of a class, including cardinality, minimum cardinality, and maximum cardinality
5. Qualified cardinality restrictions, which specify, for a given class, the number of property values coming from a particular range class—exact, minimum, or maximum
6. Equivalence classes expressing necessary and sufficient conditions for class membership
7. Union and intersection of classes
8. Disjointedness of classes
9. Two instances with different IRIs are known to actually be the same individuals or are known to not be the same.

These additional language constructs expand significantly what can be said in the language about a particular domain. They also expand the number of well-founded logical transformations that a particular model-theoretic semantics provides for a model, thereby extending the opportunity for inference of useful conclusions.

The Semantic Web technology stack defined by the World Wide Web Consortium (W3C) illustrates a rule language in relation to RDF and OWL. A rule language enables the capture of knowledge which may not be easily represented in logical axioms. As an example, consider the *age* property in our semantic model example. A *Person*'s *age* changes, so putting it into a model as an assertion would mean that the model would need to be updated regularly. However, birth dates do not change, so one might define a property *dateOfBirth* with domain *Person* and range *xsd:date*. Then one could add a rule stating that "if *x* is a *Person* and *x* has *dateOfBirth y* then *x* has *age minus*(*now*(), *y*)" where *minus* and *now* are rule built-ins. Many rule languages support the concept of small snippets of procedural code, called built-ins, that perform calculations, e.g., *now* and *minus* (get the current date–time; take the difference between two dates). There are a number of rule languages available, some of which have well-behaved and decidable semantics. For example, the Semantic Web Rule Language (SWRL) has model-theoretic semantics that are a straightforward extension of OWL semantics (SWRL: A Semantic Web Rule Language Combining OWL and RuleML 2004). The axioms of the model are extended to include the rules, and the rules become specifications of additional truth-preserving transformations.

Both the RDF and OWL languages have various syntactic representations ranging from XML-based to a functional syntax. The Manchester syntax is one of the more human-readable syntaxes (OWL 2 Web Ontology Language Manchester Syntax (Second Edition) 2012). The Semantic Application Design Language (SADL) is a controlled English representation of both OWL models and rules (Crapo A.

W., Semantic Application Design Language (SADL) 2014) (Crapo & Moitra, Towards a Unified English-Like Representaton of Semantic Models, Data, and Graph Patterns for Subject Matter Experts 2013). JavaScript Object Notation (JSON) has become a popular data serialization, and RDF data can be represented in a format called JSON-LD, the LD referring to linked data (A JSON-based Serialization for Linked Data 2014).

Evaluation of Semantic Web Languages Against Criteria

Taking Semantic Web languages as our base case, let us consider how well they meet our evaluation criteria:

1. *How well does the representation capture a domain's data and knowledge in a form compatible with the mental models of, and therefore more easily understood by, SMEs and those with basic domain knowledge?* An ontology language allows the concepts of a domain to be defined in terms of the criteria for class membership, expressed as class axioms, and the kinds of properties and relationships that members of a class can/must have. This model in essence creates a domain-specific language (DSL) that can align arbitrarily well with the shared mental models of domain experts. The model can also be instructive to those with enough foundational knowledge of the domain to be able to learn from the model.
2. *How easily can information be retrieved by the same users?* Since an RDF or OWL model is a directed graph and the data is self-describing in terms of the concepts defined in the model, queries used for data retrieval can be expressed quite easily in the DSL of the model. The Simple Protocol and RDF Query Language (SPARQL) graph query language is a powerful language for retrieving data. However, SPARQL does not allow easy retrieval of information from the logical axioms of an OWL model. The graph structures capturing these axioms are quite complex (although not necessarily so when expressed in predicate logic), and a great deal of knowledge of these structures is required to compose a query. There is a significant need for a query language that makes "meta-model" querying simple and intuitive. Several have been proposed including SPARQL-DL (SPARQL-DL API 2014).
3. *How expressive is the representation for enabling computation of new information from existing information, that is, enabling reasoning and inference?* The expressiveness of Semantic Web languages range from the simplicity of RDF to the complexity of OWL Full. Different dialects of OWL have different model theories, and hence are able to create different entailments. OWL dialects that limit their expressiveness to a subset of first-order logic, and so are not as expressive as some knowledge representations that allow fragments of higher-order and/or modal logic, are decidable. More expressive dialects may be well behaved for particular problem domains. OWL is not well suited for temporal modeling and reasoning.

4. *How well does the representation support interoperability between disparate information systems?* Several capabilities of Semantic Web languages cause it to excel in this category. This is not surprising since the purpose of the languages is to support sharing and leveraging information in the wilds of the World Wide Web. The use of XML namespaces and IRIs to establish identity is significant in achieving interoperability. Other approaches to identity are plagued with issues of global uniqueness on the one hand and issues of connecting identity between local information stores with different identifiers on the other. The inevitability of different knowledge bases creating different IRIs for the same thing is addressed by the OWL *sameAs* and *differentFrom* properties for relating IRIs. An example of the kind of interoperability enabled by Semantic Web technologies is the semantic data framework used to allow amateur astronomers worldwide to share their observations (Fox et al. 2009).
5. *Can data expressed in the representation "speak" to minds and machines with a meaning shared by both?* Meaning is emergent from structure. To the degree that an ontology language allows classes, properties, and individuals to be defined by the creation of structures that are aligned with the mental models of SMEs, the self-describing data expressed in terms of the ontology will speak to minds. We find OWL to be better than other mainstream representations for allowing the data to speak to the SMEs and not just to the programmers and database administrators. As far as speaking to machines is concerned, for representations without a model-theoretic semantics all "meaning" is generated only in the application code. Data using a model-theoretic representation speaks to machines implicitly in the sense that logical entailments are realized without any specialized application development.

5.2 How Are Semantic Web Languages Different?

Comparison with and Evaluation of XML

We used XML as an illustrative representation in the description of semantic technology above because it is well suited to highlighting the difference between syntax and semantics. XML is an acronym for eXtensible Markup Language. XML is significant in the evolution of data representation because it achieves syntactic interoperability. All compliant XML documents can be consumed by a generic (domain-independent) processor, also commonly called an XML parser. This is because XML specifies the syntax (format) of the document or data.

The syntax of XML is similar to that of other markup languages such as HTML. The basic unit is an "element" which is everything from a start "tag" to its corresponding end tag, inclusive, e.g., *<name>James Bond</name>*. Between the start and end tags, an element may contain (1) other elements, (2) text, and/or (3) attributes. Because an element can contain other elements, an XML document can always be viewed as a tree structure whose nodes are elements. Attributes are included in the start tag. Unlike HTML, XML tags must always be closed. An element which only contains attributes can be written as an empty element tag, e.g., *<product prodid="12345"/>*.

XML is eXtensible because one may define new tags at will. In fact, there are no predefined tags. Once the XML is parsed, an application may do as it wishes with the data. XML says nothing about how applications interpret that data. Thus, all of the semantics or meaning attributed to the data is provided by the application. Even so, data represented in XML may be thought of as self-describing in that the tags ostensibly describe the data to some application, and the tags are stored with the data. This self-description, along with syntactic interoperability, is a major contribution of XML.

XML schema languages allow one to exert greater control over the content of a document. Two such languages are document type definition (DTD) and XML schema. XML schema languages can be used to constrain the set of elements that may be used in a document, the attributes that can be applied to them, the values those attributes can have, the order in which they may appear, and the sub-elements which an element in the document may/must contain.

While the constraints that can be expressed in an XML schema language have some overlap with the axioms of a logic-based language such as OWL, e.g., one might think of constraining the number of sub-elements as being a cardinality restriction; by and large these constraints are more about constraining the syntax of the data than they are about creating the rich semantic structure from which meaning can emerge. To illustrate this difference, RDF and OWL can be expressed in conformant XML and one might capture the structural requirements of valid RDF/XML in a schema language. However, such a schema description would tell us the syntactic requirements of the representation (RDF/XML), and would tell us nothing about the concepts and relationships of any domain model, which is the semantics of interest to this discussion. It would be quite laborious to express the logical axioms of a complex intensional class definition in an XML schema language.

To finish our discussion of XML, we relate it to our evaluation criteria:

1. *How well does the representation capture a domain's data and knowledge in a form compatible with the mental models of, and therefore more easily understood by, SMEs and those with basic domain knowledge?* As the illustrative example in the previous section demonstrated, to the degree that the tags of an XML document are drawn from the vocabulary of the human reader, XML can be reasonably easy to read once the person is familiar with XML syntax. For those not intimately familiar with the concepts of the domain, which presumably have been used as tags, learning from the model is restricted to examining documents and associated XML schema to try to deduce the meaning of the terms.
2. *How easily can information be retrieved by the same users?* Information retrieval from XML documents requires a high level of knowledge of the schema of the document—of where particular pieces of information of interest are found in the document element tree. This is somewhat alleviated by wild cards in query languages such as Xpath which essentially allow the questioner to say "go down

any number of levels until you find this tag or attribute." However, the queries will always be more closely related to the syntax of XML than to the conceptualization of the domain.

3. *How expressive is the representation for enabling computation of new information from existing information, that is, enabling reasoning and inference?* The most that can be done with domain-independent processing of an XML document is to verify that (a) it is conformant XML and (b) it is valid according to some XML schema. Beyond that, any computation of derived information must be done in application code, eXtensible Style Sheet Language Transformations (XSLT) being an example of domain-dependent transformation. Thus, all semantic comprehension exists outside of the domain-independent XML-related code.
4. *How well does the representation support interoperability between disparate information systems?* If multiple systems have been programmed to use XML data conforming to a particular schema, then data will be interoperable between the systems. However, for the data to be interoperable with an additional system, that system must either borrow from and reuse code from an existing compatible system or have its own newly created code base that captures the meaning of the data.
5. *Can data expressed in the representation "speak" to minds and machines with a meaning shared by both?* XML data can speak to the mind of an XML and domain-savvy human, but since it cannot speak to a machine other than within the application code, the degree of shared understanding will basically depend on whether the person understands the application code. Understanding someone else's application code depends upon knowledge of the coding language and the clarity of architecting, programming, and commenting exercised by the coder.

Comparison and Evaluation of Object-Oriented Languages

In many ways, object-oriented (O-O) languages align quite well with the foundational concepts of modeling espoused in this chapter. The concept of class aligns with sets in mathematics and with the less formal but intuitive groupings that people naturally employ. Most O-O languages support some form of inheritance, which is aligned with class–subclass hierarchies. Instances of a class are instantiated in computer memory, and may have external representation through object serialization. Interclass relationships and class properties can be represented in the fields of a class.

Perhaps the real strength of O-O is its ability to capture behavior in class methods and encapsulate that behavior within the class. Behaviors can be inherited and overridden in subclasses. All of this helps to make O-O code more modular and maintainable. When exposed in a visualization other than the code, e.g., UML class diagrams, class hierarchies and interclass relationships can be quite understandable to a nonprogrammer domain expert because the classes and relationships tend to be expressed in domain terms.

One of the clear differences between most object-oriented languages and an ontology language such as OWL has to do with relationships. In a language such as Java, relationships can never be defined except within a class—they are not first-class citizens with classes. While the type of a field expresses the range of the property, the domain of the property is implicitly the class containing the field and, of course, its subclasses. If another class has a field by the same name, is it the same property or a different property? The answer is arguably a different property, as it can have a different range and can only be the "same" in the coder's mind. For example, people and businesses can have tax identification numbers, but to make a shared tax ID property, one must place the property in a superclass of both person and business. If tax ID is added to an existing model, the amount of change required to add it as a single property in both classes can be extensive. What is missing is any clear way to identify properties beyond the scope of the class.

It is well known that O-O languages are good at organizing code into class methods. This is fortuitous since a lot of code will often be required. While efforts have been made to define a model-theoretic semantics for particular O-O languages (Gunter and Mitchell 1994), it has proven difficult to do so. Therefore, the semantic interpretation of the data is left to the programmer to embed in the code. There are no logical entailments beyond the inheritance of meta-model (i.e., fields, field types, and methods) by subclasses and inheritance of class constants by instances of a class. This achieves code reuse by subclasses but the semantics remains embodied in the code.

Here is our assessment of O-O languages against our criteria:

1. *How well does the representation capture a domain's data and knowledge in a form compatible with the mental models of, and therefore more easily understood by, SMEs and those with basic domain knowledge?* The class structure of O-O languages can align quite well with the mental models of domain experts. The difficulty, rather, is in making O-O models available to nonprogrammers for their inspection and review. Looking at code is not a viable alternative for most SMEs unless they are also programmers in that language. For programmers, integrated development environments often provide a rich set of tools for viewing class hierarchies, for jumping from one class definition to a related one, etc. If UML models were created in the beginning, and the code was generated from the UML, and the UML and the code have remained synchronized, UML diagrams can be an effective way to expose the model to a broader audience. It is also possible to generate at least some UML models, e.g., class diagrams, from existing code in some languages.
2. *How easily can information be retrieved by the same users?* Retrieving information from an actual O-O instance model, in-memory or serialized, is not a common practice nor is there any standard query language for doing so. Programming environments will normally support searching over the meta-model (class definitions) to find classes or properties, but these tools generally do not expose an instance data view and are only accessible to programmers. This difficulty has often been addressed through a combination of O-O programming and

relational database storage, allowing data to be queried in the relational store but computed over in the O-O methods.

3. *How expressive is the representation for enabling computation of new information from existing information, that is, enabling reasoning and inference?* The computation of new information from existing information is almost entirely up to the class methods. The question of the well foundedness of such computations is one of creating validation tests, and then verifying that the code passes those tests. There is not, in general, any formal way of verifying that the tests themselves are correct, consistent, complete, etc.
4. *How well does the representation support interoperability between disparate information systems?* The hope at the outset of the object-oriented age was that classes would be reused at a scale that would enable interoperability because different systems and applications would be using the same underlying O-O code base. For whatever reason, this vision was largely unrealized. One might say that it has partially come to pass in that the code to parse syntactic representations, e.g., XML parsers, and syntactic transformations has been created using O-O languages, and these classes are often used and reused as building blocks for higher-level semantic languages like OWL.
5. *Can data expressed in the representation "speak" to minds and machines with a meaning shared by both?* Like XML, data captured in O-O class instances has meaning within the applications that use the data, but it is difficult to tell whether that meaning is shared by domain experts because portions of the model, especially the code in the methods, can be quite opaque to SMEs.

Relational Revisited: Comparison and Evaluation

This section continues the discussion of relational technology started in the earlier chapter section *The Data World Today*. Relational technology is founded on first-order predicate logic. An n-ary relation, commonly called a table, is a set of tuples $(d_1, d_2, \ldots, d_k)$ where each element of a tuple is a member of a corresponding data domain, commonly called a data type. The combination of an element's name and its domain is commonly called an attribute or column. Under the closed-world assumption, which essentially means that what is not known to be true is assumed to be false, the n-ary relation is the extension of an n-ary predicate. In other words, the tuples are those sets of values which make the predicate true. Database constraints, such as referential integrity, can be expressed in first-order formulas. Similarly, a relational query language is possible which consists of statements in first-order logic. Relational technology has in theory, although implementations may deviate, a well-founded logical model.

So, one might ask, what does relational technology lack? One of the key issues of relational technology is the lack of a good "object model" applying set theory to create partial ordering of classes into class hierarchies and thus enabling inheritance. There have been various attempts to bring an object model into the relational world but there has not been consensus on how to do so. Some feel that many current relational implementations have deviated unforgivably from their logic foundations (Date and Darwen 2006). A fundamental question to be answered in incorporating an object model is what, in the relational model, corresponds to a set or class? In fact, one might even ask of a relation's primary key (the combination of tuple elements which guarantee a unique matching tuple), what does the key identify, other than the tuple? Interestingly, OWL 2 introduces the concept of a key on a class definition, but here it is clear that the combination of key properties is a unique identifier of an instance of the class, differentiating that instance from all others. Keys in the OWL sense are particularly useful to establish identity in the case of blank nodes. This is distinctly different from the system-generated key column in some relational implementations, as this surrogate key does not identify a set of characteristics of an instance which are sufficient to uniquely identify the instance.

While solidly built on first-order predicate logic, this lack of alignment of the relational model with set theory, even an informal set theory, creates a certain amount of dissonance between the model constituted by the relational schema and the mental models of SMEs. Put another way, relational schemas seem sometimes to be more about the structure of the database than about the concepts of the domain. A domain expert looking at a relational database will recognize domain concepts in the names of tables, views, and columns, but the interpretation of the semantics of the data will depend largely upon the viewer's mental model. Like XML, the semantics is provided by the human viewer of the data or by the application code using the data; there is not enough domain structure to enable meaning to emerge from the model itself.

To illustrate this point, consider two examples: the often-used supplier parts database and a continuation of the family model theme used in the section *What We Mean by Semantics*. First, consider the supplier parts database. This example is used, in various flavors, to provide an introduction to relational technology. We examine some difficulties in deriving a consistent meaning from the simple schema to illustrate how and why relational models often do not reflect a semantics shared with people. Of the supplier part database instantiations available, we choose one with the following table definitions, where *S* is the supplier table, *P* is the parts table, and *SP* is the shipment table, a junction table capturing the relationship between suppliers and parts. The example can be expressed in Tutorial Dee syntax (Cunningham and Cunningham, Inc. 2014).

```
VAR S REAL RELATION {S# CHAR, SNAME CHAR, STATUS INTEGER, CITY CHAR} KEY{S#};

VAR P REAL RELATION {P# CHAR, PNAME CHAR, COLOR CHAR, WEIGHT RATIONAL, CITY CHAR} KEY{P#};

VAR SP REAL RELATION {S# CHAR, P# CHAR, QTY INTEGER} KEY{S#, P#};
```

We assume referential integrity constraints enforcing that for every tuple in *SP*, *S#* shall be found in *S* and *P#* shall be found in *P*.

A number of issues arise in determining the semantics implied by this simple relational model. Consider the supplier table *S*. Since *S#* is the primary key, it must be unique to each supplier, and there should be only one tuple in this table with this key. The database schema therefore restricts each supplier to a single *SNAME* and a single *CITY*. To accommodate the possibility that a supplier might be known by other names, e.g., *GE* and *General Electric*, we would need to either expand the number of columns *(SNAME1, SNAME2,…)* or create a new table, *Supplier Name*, with composite primary key *{S#, SNAME}*. The first solution has the obvious disadvantage that the number of possible names is fixed by the number of columns devoted to names, and any supplier with fewer names would have one or more *NULL* values in the sname columns. The second solution creates a more complex schema structure that must be understood by a user in order to query for a supplier's name. Since a supplier may have presence in more than one *CITY*, the same solutions present themselves, and if a separate *Supplier City* table is created, only *S#* and *STATUS* are left in the original *S* table. The solution of adding columns is illustrative of one of the significant differences between a graph model and the relational model. In a graph model, when a subject does not have a property value (e.g., a second *SNAME*), it is simply absent from the graph. In the relational model, a NULL must be placed in the table—a value indicating that there is no value.

Now let us consider the *Parts* table *P*. What does the primary key, *P#*, uniquely identify? Initially, one might think that it is the unique identifier of a part. However, the fact that there is a quantity *QTY* in the *SP* table whose value presumably can be greater than one tells us that this cannot be the case. (The unique identifier of individual parts is usually called a serial number.) After consideration, one might conclude that since different suppliers can provide parts with the same *P#, P#* identifies the set of all parts which are interchangeable in their application and are possibly not identifiable individually other than by physical possession, e.g., two nuts of the same size, material, and thread. Parts whose value does not warrant the effort of tracking them individually are sometimes called fungible. But this conclusion has a problem: the *CITY* column. Different suppliers can provide parts with the same *P#*, and different suppliers can be in different cities but there can only be one tuple in *P* for a given *P#* and so a given *P#* can only have one associated *CITY* value. So how can all parts with the same *P#* be in the same *CITY*? Perhaps, what is meant by a tuple in *P* is not all of the parts with the same *P#*, but rather the subset of parts with a given *P#* which were shipped in a particular shipment identified by a tuple in *SP*. And, since the primary key of *SP* is *{S#,P#}*, a given supplier could only ship parts with a given *P#* once.

The point we are making is that although one can surely create a valid relational model for this domain, the relational model leads us to think about the data schema and not about the domain being modeled. The resulting models are not likely to be very well aligned with the way SMEs think about their domain. Nor are users of the data empowered to explore and query the data using domain terms. They will only be able to retrieve data if they have a clear understanding of the relational structure or, which is more often the case, they have a database expert in the IT department create a query for them. The lack of a set-theoretic foundation leads to schema-centric models and is, in our opinion, a significant disadvantage. By contrast, relaxing a restriction in a semantic model, such as removing a functional or cardinality restriction so that more names or locations are allowed, would have essentially no consequence to the model. All of the existing data would still be valid. If the change was in the direction of greater restriction, e.g., to allow only one name or location for a particular domain class, the axiom would be added, no data would be lost, and a logical check of the model would reveal which data was no longer consistent with the model. For these kinds of reasons, we sometimes refer to semantic models as flexible data models.

Now, let us look at the family model used above to illustrate the elements of a semantic model, and ask how this model might be represented in a relational database. One might begin with a person table *P* and a child table *C*.

VAR P REAL RELATION {PID CHAR, NAME CHAR, GENDER CHAR, AGE INTEGER} KEY{PID};

VAR C REAL RELATION {PID CHAR, CPID CHAR} KEY{PID, CPID};

Referential integrity would require that each *PID* and each *CPID* in *C* correspond to a *PID* in *P*. How would we capture the class equivalence between the *Man* class, a subclass of *Person*, and the set of all instances of *Person* with *gender Male*? How would we capture that the *child* relationship with domain *Person* and range *Person* has a sub-property *son* with range *Man*? How could we define role classes such as *Mother, Son, Uncle*, and *Ancestor*? And if we did define them what would be the mechanism for inferring these class memberships and relationships such as *father* (for all *x*, for all *y*, if *child (x, y)* and *gender (x, Male)* then *father (x, y)*)?

For most relational implementations, these questions can have only two answers. Either construct queries in a first-order language, SQL being a proxy for such, or write scripts in a first-order language, PL/SQL being a proxy for such. The first approach has the disadvantage that each user of the database must either re-implement the query or recognize it in a repository of queries as the one that will draw the appropriate inferences. Either of these choices requires that the user have a deep knowledge of the relational schema of the database. The second approach of writing scripts can certainly create and populate tables or views that identify the inferred "types" and relationships. The skillset and mental processes required to write such scripts would certainly be different from the creation of appropriate logical axioms. Using the inferred tables or views would bring us back to our primary complaint. To retrieve the inferences, one would need to have a deep understanding of the relational schema generated by the scripts, much of which would not be expert knowledge of the domain but specific knowledge about this particular schema.

In terms of our criteria of success, here is how we evaluate relational technology:

1. *How well does the representation capture a domain's data and knowledge in a form compatible with the mental models of, and therefore more easily understood by, SMEs and those with basic domain knowledge?* The relational model is lacking an "object model." Informal set theory seems to play a very significant role in the way humans perceive and make sense of the world. The lack of a

partial ordering of sets in a class hierarchy, with inheritance, makes it difficult to create relational models that seem to be about the domain and not about the relational schema. Users are forced to create mental mappings to their own mental models and to do so without the confirming feedback that a good semantic model might provide.

2. *How easily can information be retrieved by the same users?* Since the queries are expressed in terms of table and column names, as well as relational operations between tables, e.g., join, and these names are not well aligned with class and property definitions, querying becomes more about understanding the database schema than about understanding the concepts of the domain.
3. *How expressive is the representation for enabling computation of new information from existing information, that is, enabling reasoning and inference?* The relational model is well-founded in logic, theoretically if not in some implementations. However, many well-founded entailments that one might expect come from the semantics of set theory and axiomatic class definitions, which are missing from the relational model. Hence, computation is largely the task of scripts (i.e., PL/SQL), queries (i.e., SQL), or applications using the data. Each of these requires a deep understanding of the schema as well as how it maps to the concepts of the domain.
4. *How well does the representation support interoperability between disparate information systems?* Because the relational models are logically well founded, it is possible to create mappings between compatible relational schemas. Exactly what compatible means may be difficult to say. Since the schema tends to not align with a SME's domain model, determining if two schemas are compatible involves understanding the domain and how each schema maps to that domain.
5. *Can data expressed in the representation "speak" to minds and machines with a meaning shared by both?* Buying into a relational schema is, in some sense, comparable to adopting an ontology. However, the relational schema does not have the implicit benefit of making the data self-describing, and thereby meaningful to any man or machine with a compatible mental model or ontology. The ability to speak is hampered by the impossibility of creating a schema isomorphic with the mind's mental model.

Comparison and Evaluation of UML with Notes on MOF, eMOF, ODM, and SMOF

Considering the prevalence of UML as a modeling language, it seems appropriate to describe briefly how it compares with other semantic languages. UML was born of a desire to do model-driven code generation in O-O languages. Since its inception, UML's foundation and the foundation of modeling, in general, have evolved significantly. The Object Management Group (OMG) has done ground-breaking work in creating a modeling framework, resulting in the creation of the meta-object facility (MOF) and essential MOF (eMOF) (Object Management Group 2014). As interest in ontology languages has increased, OMG has sought to understand how MOF and eMOF are similar to and different from ontology language newcomers like OWL and CL. One outcome is the OMG standard Ontology Definition Metamodel (ODM) for "Model-Driven Architecture (MDA)-based software engineering" (Object Management Group 2007). ODM claims among its benefits "Options in the level of expressivity, complexity, and form available for designing and implementing conceptual models, ranging from familiar UML and ER methodologies to formal ontologies represented in description logics or first order logic" and "Grounding in formal logic, through standards-based, model-theoretic semantics for the knowledge representation languages supported, sufficient to enable reasoning engines to understand, validate, and apply ontologies developed using the ODM"(Ibid). It also promised to be "The basis for a family of specifications that marry MDA and Semantic Web technologies to support semantic web services, ontology and policy-based communications and interoperability, and declarative, policy-based applications," thereby "mov[ing] UML from a model of an O-O language toward a well-founded modeling framework" (Ibid). This promise is at least partially fulfilled in the more recent semantic MOF (SMOF), which recognizes that MOF "suffers from the same structural rigidity as many object-oriented programming systems, lacking the ability to classify objects by multiple metaclasses, the inability to dynamically reclassify objects without interrupting the object lifecycle or altering the object's identity, and a too constrained view on generalization and properties" (Object Modeling Group 2013). In other words, there is a convergence of the O-O modeling and ontology communities with potential benefits for both. UML is becoming more well founded and ontology development is benefiting from new tools and methodologies and from the ability to move knowledge into and out of ontology languages.

UML is mainly about the meta-model, and is not generally used to model instance data. Some efforts have been made to create "computable UML" (Milicev 2009), which would of necessity include instance data if the computation envisioned was scenario specific, not just entailments over the meta-model. However, for the present, UML does not rise to the level of a logically well-founded language.

Here's how we see UML stacking up against the criteria:

1. *How well does the representation capture a domain's data and knowledge in a form compatible with the mental models of, and therefore more easily understood by, SMEs and those with basic domain knowledge?* UML does a very good job of capturing meta-models that are understandable by SMEs. This is one of the reasons it was developed. What it does not enable is looking at models of scenario or instance data and making this model transparent to domain experts.
2. *How easily can information be retrieved by the same users?* We are unaware of a query language for UML. Most UML modeling environments will have search capabilities that will help developers find elements of a UML meta-model. Since UML does not generally capture instance data, query and search are mainly restricted to the meta-model.
3. *How expressive is the representation for enabling computation of new information from existing information, that is, enabling reasoning and inference?*

UML does not, generally, support inferencing of well-founded entailments over instance data. A certain level of inferencing over the meta-model may be implemented by a particular UML development environment but, lacking a complete model theory and proof theory, computational results are not uniform or provably truth preserving.

4. *How well does the representation support interoperability between disparate information systems?* If two systems shared the same UML model of the domain, and the generated O-O code were primarily data containing and not augmented with incompatible methods, significant interoperability could be achieved. However, the UML model would be of limited use in achieving interoperability with another existing system. Most UML-modeled applications depend upon a repository, e.g., relational database, to provide data across multiple applications.
5. *Can data expressed in the representation "speak" to minds and machines with a meaning shared by both?* To the degree that the UML model is sensible to domain experts, and the model is also used to generate O-O code, there is a sharing of at least the basic class structure between man and machine. However, the semantic interpretation of the model from a truth-preserving transformation or statement denotation perspective would be entirely the obligation of the method code, which usually is not entirely derived from the UML model.

Summary of Representations

Table 5.2 contains a comparative summary of our evaluation of representations/languages discussed in this chapter. In place of a descriptive analysis of each, a letter grade is provided with "A" being complete fulfillment of the criteria and "F" being complete failure to meet the objective.

Our intent has not been an exhaustive comparison with all possible representations but rather a discussion of a cross section of popular modeling paradigms. We

Table 5.2 Comparison and evaluation of selected representation technologies

Criteria/Technology	XML	O-O	Relational	UML	Semantic
1. Compatibility with SMEs mental models	C	B	C	B	A
2. Easy retrieval in domain terms	C	D	D	F	B
3. Well-founded reasoning and inference	F	D	C	D	B
4. Interoperability	D	F	C	F	A
5. "Speaks" to minds and machines with shared meaning	F	D	D	D	A

This table only shows an overall grade for each considered technology for each evaluation criterion. For a more detailed discussion, see the evaluation/comparison by criterion of each technology in the section discussing that technology close this part of the chapter by sharing briefly some observations about gaps that are not addressed by current ontology languages and some opinions about how to best apply what does exist. First, there are many fine computational mechanisms that seem way beyond the capability of semantics to efficiently capture and implement at this time, e.g., regression analysis of large datasets. It is not feasible at present to move these computations "into" a semantic model. What we think semantics adds is the ability to model what the analytic does, what kind of data it requires as input, and the meaning of the output. In other words, semantic models can be used at the higher level, and analytic programs can return their results to that higher level for interpretation, sharing, and use by humans and other computational mechanisms, model theoretic, or otherwise.

Our second observation is that the mainstream ontology languages of today are not well suited to temporal modeling, either of large quantities of time series data or of the time domain in general. For example, modeling of a real-time operating system in OWL is a stretch. As noted in a prior section, temporal modeling can be done using modal logic. Clearly, ontology languages like OWL are not the solution to every modeling problem. However, they are good at capturing certain kinds of information about the domain and making it available to models that do handle temporal constraints more gracefully. Our vision is a semantic model foundation on which lots of other models and algorithms can build. This has the advantage of providing a cross-enterprise, possibly cross-industry, shared domain model for understanding and interoperability while allowing many contributors to add valuable data and insights as they are able. The semantic foundation will make these contributions understandable and valuable to both minds and machines.

5.3 Semantics in Real-World Solutions

In this section, we describe five real-world applications using the semantic approaches we described earlier. Each application will serve to highlight one of our five evaluation criteria, which we abbreviate as *common understanding, information availability, inference, interoperability*, and *data speaks*, respectively. In all the examples, we have deployed a common stack of semantic technologies, favoring open source solutions where possible, and always preferring standards, like OWL, over proprietary representations.

Model-Driven Equipment Overhaul: Common Understanding

One of the early motivations for using semantics within our team was to enable information scientists to more easily deliver technology solutions to our maintenance engineers. Within the industrial maintenance practice, engineers develop proprietary guidelines and rules, as well as business preferences, describing how to overhaul a piece of equipment. Expert systems are particularly applicable in this domain as they allow multiple competing objectives to be reasoned about to provide a set of best possible solutions. However, the knowledge representations useful to capture domain knowledge are not always the same representations that best allow non logicians or non information scientists to work collaboratively with the software. This is particularly important when one considers the maintenance of the knowledge bases themselves—if the domain experts do not easily understand the data and knowledge representation, it is very likely that mistakes will be made or the system will not be kept current, both of which deficiencies lead to a loss of value that the knowledge-based system can deliver.

To overcome these challenges, we developed a solution that describes the equipment and maintenance practices in OWL plus rules. We found the available tools to be more suitable for ontologists and computer scientists than engineers, so we began to experiment with a more English-like DSL for expressing the semantic models. The result was our aforementioned SADL technology which currently has OWL 1 expressivity, plus qualified cardinality, and incorporates semantic highlighting, hyperlinking, etc. A valid SADL model is translated to and saved as OWL whenever the SADL model is saved. The rules are translated to the target rule language, e.g., Jena Rules. The SADL development environment also supports a compatible query language, tests and test suites, warnings about constructs likely to have unintended entailments, and rule debugging aids. We also created a template that essentially creates a few-column representation of overhaul knowledge that is easy for engineers to add to and maintain and which is translated into rules in the SADL language, from which they are automatically translated to the target rule language.

As we developed the SADL language, we used engineers in a GE business as a litmus test. We would construct a model of some portion of their domain in SADL and then show them the model in a collaborative teleconference/desktop sharing session. If they asked us what the model meant, we had failed. If they immediately began to tell us what was wrong with or missing from the model, we counted it as a success. We wanted the modeling language to be transparent to them so that they could focus on the model. We were astonished at the degree to which SADL was a success in this respect. We felt that we were indeed achieving a *Common Understanding* between mind and machine. Over time, we have added various other technologies to enable knowledge delivery and application integration. The engineers across the maintenance organization have been able to create new equipment models, create new overhaul rules and guidelines, and maintain both.

Data Provenance: Information Availability

In joint work with Lockheed Martin, we have developed models of data flow, data provenance (DP), and information assurance properties of data in multilevel secure environments (Moitra et al. 2010) (Dill et al. 2012). DP includes information sources, ownership, processing details, and other attributes. Information assurance attributes include confidentiality, authenticity, integrity, and non-repudiation. At each point in information flow or storage there is opportunity for the data to be compromised in ways reflected by these assurance attributes. When information flows across a security-level boundary, information redaction may be necessary. For example, the source of the data may be "secret," and so when passing to a "non-secret" security level, the source as well as part of the data itself may need to be removed. Our objective was to model the workflow for transmission of simple messages and complex messages (those having attachments) in such a way that at each point the information assurance attributes of the message could be summarized in a figure of merit (FoM) and displayed to the user with drilldown capability. The FoM allowed recipients of messages in security levels lower than their origination to have an indication of the messages' information assurance attributes in the absence of complete provenance.

Each time a message is moved between agents, systems, or processes, a DP record is created with two parts, one from the sender and one from the receiver. The provenance information may be stored and/or sent in parallel with the message. Subjective logic, which distributes unity between belief, disbelief, and unknown (Josang 2013), is used as the algebra for combining assurance attributes from multiple DP records. The semantic model captures the conceptualization of the problem of information assurance in terms familiar to those working in the field. Retrieval of high-level information, e.g., FoM, and drilldown to the details of how the summarized information is computed is relatively easy and can be accomplished in domain terms. The flow of messages with the provenance at each point can easily be viewed graphically. A key feature of the work is the relative ease with which the information can be retrieved after analysis of a message flow scenario.

Early Manufacturability: Inference

Traditional manufacturing separates the function of product design from the actual manufacturing itself. This separation can lead to long-cycle iterations and increased cost as a completed design passed on to manufacturing is discovered to be difficult or impossible to manufacture and is returned to engineering for redesign. A solution to this problem is to provide feedback on manufacturability to design engineers during the design process. We have developed semantic models of solid models and design and manufacturing processes that serve as a foundation for manufacturability rules and enable manufacturability analysis of the design to be fed back to the designer in the computer-aided design (CAD) platform (Rangarajan et al. 2013).

Inference occurs on multiple levels to provide relevant and timely feedback to design engineers. The models have a rich class hierarchy of features, surfaces, tools, and processes, which, along with their properties, characterize the solid models of the parts under design and characterize how they might be produced. Manufacturability rules can be expressed over classes at various levels in this hierarchy, with the implications of rules at higher levels being inherited by subclasses at lower levels. This allows parsimony of expression, which is essential to the simplicity and maintainability of the knowledge base. Where more convenient or more efficient, low-level calculation of geometric characteristics of features, e.g., dimensions, angles, etc., are performed in the procedural code that links the knowledge base and reasoner to the CAD system. The results of these computations are inserted into the semantic model. This illustrates the utility of the semantic model as the recipient of information computed outside of the model, and as the source of the new information integrated with the old, now available for additional reasoning and/or for easy access by other systems and users.

Smart Grid: Interoperability

Traditional energy generation, transmission, and distribution techniques and systems are increasingly inadequate to meet the challenges of a world with increasing renewable energy sources, such as wind and solar, and increasing incentives to use energy more efficiently. The term "smart grid" has been coined as the name of a new generation of technologies that use digital communication systems and more flexible electricity generation and delivery systems to meet these demands. One of the challenges of smart grid is the level of interoperability required between system components, both traditional grid components, and elements of the new communication and control infrastructure. It is our belief that semantic technology will play a significant role in achieving the required model transparency and data interoperability (Crapo et al. 2009).

The International Electrotechnical Commission (IEC) Standard 61970: Common Information Model (CIM)/Energy Management and related standards have been essential to the evolution and stability of the electric utility industry. The CIM, which has been in the making for years, is a semantic model captured in UML as well as a textual standard. Only recently has Semantic Web technology offered an alternative for modeling the domain. In our work, applying OWL models and rules to electric utility data, we found that utility companies already exchange descriptions of distribution system networks as RDF using translations of the CIM into RDF and OWL as the metadata. An obvious reason for their apparently early adoption of RDF is that the UML models of the CIM, while very useful in defining classes and relationships in the meta-model and making these available via visualizations, are not well suited to marking up actual data about physical systems so that it is self-describing, easily queried, and easily exchanged with other information systems. While we found the current OWL translations of the CIM to be lacking in important respects (Crapo et al. 2010), we found such models to be extensible, executable, useful for data validation, and valuable as a foundation for additional logic captured in rules. We identified several ways in which the CIM can benefit from a transition to a set of compatible ontologies. And we see the future smart grid as just one instantiation of the Internet of Things, or the Industrial Internet, where minds and machines are connected and data is interoperable because it is expressed in semantic representations (Crapo, Piasecki, & Wang, The Smart Grid as a Semantically Enabled Internet of Things 2011).

Data Science: Data Speaks

As an increasing number of sensors are placed within today's industrial processes, the volume and variety of data produced are dramatically increasing. To handle the volume of data, it is typical to use time series "historians" and "event-driven" data models to enable very fast write throughput of the data coming from the industrial SCADA systems into databases. Whereas Complex Event Processing (CEP) engines address application of rules and data mining on the high-throughput stream of data in real time, data scientists, or data analysts, must query historical data for post-analysis and identification of new insights for improving processes.

Recently, such a data science challenge was presented to us. GE's "brilliant factory" initiative envisions a new generation of manufacturing plants where a single digital thread extends from design and engineering through production, distribution, and service over the life of the product, and reaches back into the supply chain (GE 2014). Feedback from every stage in the product lifecycle, including design, engineering, production and field service, will improve the product over time. GE Energy Storage's US$ 100 million Durathon battery manufacturing facility in Schenectady, NY, is a test bed for the factory of the future (GE Energy Storage n.d.). This facility can also serve to illustrate the application of semantic technology to a real-world data science problem.

Real-time data collection on the factory floor is a challenging task, and the demands of making sure the data is reliably captured can drive the data repository schema to be more capture oriented than domain oriented. This can present a significant challenge for SMEs and data analysts who understand the factory but not necessarily the data schema. The result is that a quality engineer, for example, may need the help of a team of database experts in order to obtain data to analyze a quality issue. Getting the queries to get the data can take a considerable amount of time and slow down the process of improvement, resulting in lost opportunity.

Starting with a general semantic model of process, material, substance, consumption, production, etc., we then extended this model to the particulars of the battery manufacturing facility. This domain-specific "meta-model" was then available for query, exploration, etc. The next step was to map the existing data stores to the semantic model. In theory, this mapping can be used to import the data into a triple store or to map semantic queries to queries against the existing data repositories. In practice, the former approach is more doable with current technology and allows for better data retrieval performance. Once the data was transformed to the semantic domain, it could be retrieved in terms of domain concepts. The semantic model is a graph in which all nodes are domain concepts (processes or materials) as are all edges (relationships such as consumes or produces), and therefore one can "walk the graph" from any node "upstream" or "downstream" in domain terms, making data retrieval sensible to data consumers. For example, one might ask for the pressure and temperature of some process step, identified by type, in the manufacture of a particular cell, or the average of these values for all of the cells of a particular battery, or of all of the cells produced from a particular upstream material lot. The semantic data model allows genealogy queries like these to be completely devoid of any reference to the underlying data schema.

We expect our modeling effort for the Durathon battery plant to be applicable to other kinds of manufacturing facilities. For example, one can generate a schematic view of all of the material flows and processes involved in creating a Durathon battery with a query that contains only generic terms such as process, produces, etc. In fact, we used these schematic diagrams created directly and automatically from the battery plant models to allow plant engineers to understand and iteratively correct our models with little time and effort on their part. The factory-specific extensions of the generic manufacturing model must, of course, be constructed for the new facility. If the factory already exists, the mappings from the existing data stores to this semantic model must also be customized to the specific factory schema and the domain-specific model. If the factory is new, it may be possible to generate more domain-related data capture schemas. As we repeat this process for subsequent factories, we expect to learn how to streamline the customization process and better generalize reusable model and data-mapping components. In this way, data collected in brilliant factories will speak to those who operate them.

5.4 Conclusion

In this chapter, we have explored the foundational concepts underpinning several popular information and knowledge representation technologies and languages and compared them with Semantic Web languages. We have identified areas where we believe Semantic Web languages are superior, but we have also identified opportunities for improvement. We have argued that representations are most useful when they align with SMEs' mental models of the domain and support truth-preserving transformations leading to well-founded inference. The combination moves us in the direction of our vision for the Industrial Internet—an environment where data speaks to minds and machines alike, enabling a new level of mind–machine collaboration to better solve tomorrow's more complex problems.

References

A JSON-based Serialization for Linked Data. (2014, January 16). Retrieved from W3C: http://www.w3.org/TR/json-ld/. Accessed 23 May 2015.

Baader, F., Calvanese, D., McGuinness, D., Nardi, D., & Patel-Schneider, P. (2003). *The description logic handbook: Theory, implementation, and applications*. Cambridge: Cambridge University Press.

Bizer, C., Heath, T., & Berners-Lee, T. (2009). Linked data–the story so far. *InternationalJjournal on Semantic Web and Information Systems*, 5(3) 1–22.

Crapo, A. W. (2014, September 25). *Semantic application design language (SADL)*. Retrieved from SourceForge: http://sadl.sourceforge.net/. Accessed 23 May 2015.

Crapo, A. W., & Moitra, A. (2013). Towards a unified english-like representaton of semantic models, data, and graph patterns for subject matter experts. *International Journal of Semantic Computing, 7*(3), 215–236.
Crapo, A., Wang, X., Lizzi, J., & Larson, R. (2009). *The semantically enabled smart grid. Grid-Interop*. Denver: GridWise Architecture Council. Retrieved from http://www.gridwiseac.org/pdfs/forum_papers09/crapo.pdf. Accessed 23 May 2015.
Crapo, A., Griffith, K., Khandelwal, A., Lizzi, J., Moitra, A., & Wang, X. (2010). *Overcoming challenges using the CIM as a semantic model for energy applications. Grid-Interop 2010*. Chicago: GridWise Architecture Council. Retrieved from http://www.smartgridnews.com/artman/uploads/1/crapo_gi10.pdf. Accessed 23 May 2015.
Crapo, A., Piasecki, R., & Wang, X. (2011). *The smart grid as a semantically enabled internet of things. Grid-Interop 2011*. Phoenix: GridWise Architecture Council. Retrieved from http://www.pointview.com/data/files/3/2433/2137.pdf. Accessed 23 May 2015.
Cunningham & Cunningham, Inc. (2014, September 26). *Supplier parts database*. Retrieved from http://c2.com/cgi/wiki?SupplierPartsDatabase. Accessed 23 May 2015.
Cyc. (2014, September 25). Retrieved from Wikipedia: http://en.wikipedia.org/wiki/Cyc. Accessed 23 May 2015.
data.gov. (2014, November 15). Retrieved from Wikipedia: http://en.wikipedia.org/wiki/Data.gov. Accessed 23 May 2015.
Date, C. J., & Darwen, H. (2006). *Databases, types, and the relational model: The third Manifesto*, (3rd ed). Reading, MA: Addison-Wesley.
Dill, S. J., Barnett, B., Crapo, A., & Moitra, A. (2012). *Patent No. 8166122*. USA. Retrieved from http://google.com.ar/patents/us8166122. Accessed 23 May 2015.
Fox, P., McGuinness, D. L., West, P., Garcia, J., Benedict, J. L., & Middleton, D. (2009). Ontology-supported scientific data frameworks: The virtual solar-terrestrial observatory experience. *Computers & Geosciences, 35*(4), 724–738.
GE. (2014, February 25). *Meet your maker: The third industrial revolution will be crowdsourced and digitized*. Retrieved from GE Reports: http://www.gereports.com/post/77834521966/meet-your-maker. Accessed 23 May 2015.
GE Energy Storage. (n.d.). *Durathon media center*. Retrieved from GE Energy Storage: http://renewables.gepower.com/energy-storage.html. Accessed 23 May 2015.
Gunter, C. A., & Mitchell, J. C. (1994). *Theoretical aspects of object-oriented programming: Types, semantics, and language design*. Massachusetts: Institute of Technology.
Gurr, C. A. (1997). On the isomorphism, or lack of it, of representations. In K. Marriot, & B. Meyer (Eds.), Visual Language Theory, ISBN 928-1-14612-7240. New York: Springer Science & Business, pp. 293–305.
Gustafson, S., & Sheth, A. (2014, March). The Web of Things. Retrieved from Computing Now, Vol. 7, Number 3, IEEE Computer Society: http://www.computer.org/portal/web/computing-now/archive/march2014. Accessed 23 May 2015.
Information technology–Common Logic Standard (CL): a framework for a family of logic-based languages. (2007, October 01). Retrieved from ISO Standards Maintenance Portal: http://standards.iso.org/ittf/PubliclyAvailableStandards/c039175_ISO_IEC_24707_2007!!/span>%28E%29.zip. Accessed 23 May 2015.
Johnson-Laird, P. N. (1983). *Mental models: Towards a cognitive science of language, inference, and consciousness*. Cambridge: Harvard university Press.
Johnson-Laird, P. N. (1988). *The computer and the mind*. Cambridge: Harvard University Press.
Johnson-Laird, P. N. (1993). *Human and machine thinnking*. Hillsdale: Lawrence Erlbaum Associates.
Josang, A. (2013, February 18). *Subjective logic (Draft)*. Retrieved from http://folk.uio.no/josang/papers/subjective_logic.pdf. Accessed 23 May 2015.
Marr, D. (1982). *Vision: A computational investigation into the human representation and processing of visual information*. San Francisco: W. H. Freeman and Company.
Milicev, D. (2009). *Model-driven development with executable UML*. Indianapolis: Wiley.
Moitra, A., Barnett, B., Crapo, A., & Dill, S. J. (2010). Using data provenance to measure information assurance attributes, June 15–16. In D. McGuinness, J. Michaelis,& A. Khandel-

wal, L. Moreau(Eds.) Provenance and Annotation of Data and Process: *Third International Provenance and Annotation Workshop*. Troy, NY, June 15-16, 2010, Revises Selected Papers, Springer Science & Business Media, New York, pp 111–119.

Motik, B., Grau, B. C., Horrocks, I., Wu, Z., Fokoue, A., & Lutz, C. (2012, December 11). *OWL 2 web ontology language profiles* (2nd ed). Retrieved from W3C: http://www.w3.org/TR/owl2-profiles/. Accessed 23 May 2015.

Object Management Group. (2007). *Ontology definition metamodel*. Retrieved from http://www.omg.org/cgi-bin/doc?ptc/07-09-09.pdf. Accessed 23 May 2015.

Object Management Group. (2014, September 25). *OMG's metaobject facility*. Retrieved from http://www.omg.org/mof/. Accessed 23 May 2015.

Object Modeling Group. (2013, April 2). *MOF support for semantic structures*. Retrieved from http://www.omg.org/spec/SMOF/1.0/PDF. Accessed 23 May 2015.

OWL 2 Web Ontology Language Manchester Syntax (Second Edition). (2012, December 11). Retrieved from W3C: http://www.w3.org/TR/owl2-manchester-syntax/. Accessed 23 May 2015.

Pan, B., & Gleason, J. B. (1997). Semiotic development: Learning the meanings of words. In J. B. Gleason (Ed.), *The development of language*, (4th ed). Boston: Allyn and Bacon.

Rangarajan, A., Radhakrishnan, P., Moitra, A., Crapo, A., & Robinson, D. (2013). Manufacturability analysis and design feedback system developed using semantic framework. *18th design for manufacturing and the life cycle conference*. Portland: ASME Proceedings.

Resource Description Framework (RDF). (2014, September 25). Retrieved from W3C: http://www.w3.org/RDF/. Accessed 23 May 2015.

Sowa, J. F. (2014, September 24). *The role of logic and ontology in language and reasoning*. Retrieved from http://www.jfsowa.com/pubs/rolelog.pdf. Accessed 23 May 2015.

SPARQL-DL API. (2014, September 25). Retrieved from derivo Symantic Systems: http://www.derivo.de/en/resources/sparql-dl-api/. Accessed 23 May 2015.

SWRL: A Semantic Web Rule Language Combining OWL and RuleML. (2004, May 21). Retrieved from W3C: http://www.w3.org/Submission/SWRL/#3. Accessed 23 May 2015.

Wade, N. J., & Swanston, M. (1991). *Visual perception*. London: Routledge.

Chapter 6
Unnatural Language Processing: Characterizing the Challenges in Translating Natural Language Semantics into Ontology Semantics

Kent D. Bimson and Richard D. Hull

6.1 Introduction

In recent years, significant work has been performed on using ontologies as a basis for extracting knowledge from text and representing it as ontology knowledge structures, typically using Web Ontology Language (OWL; Bechhofer et al. 2004) or a Resource Description Framework (RDF) (RDF—Semantic Web Standards 2004) triple store. There are many good reasons for doing so, such as to capture specific knowledge from text that is needed for a domain analysis problem, to structure and normalize semantics for machine interpretation, to improve semantic search, or to share information via the Semantic Web, to name only a few.

Significant challenges arise, however, when ontology designers attempt to model the richness of natural language (NL) semantics using constrained ontology semantics (Bimson 2009), a process that leads to what we call unnatural language processing, or attempting to model the richness of NL semantics using constrained ontology semantics. Unnatural language processing is a consequence of the fact that the semantics natural to languages is not so easily or naturally represented within standard ontology representation languages, at least not without significant effort. As a result, much of the semantic content in NL text is lost in translation to RDF or OWL. Semantic extensions added to OWL 2 (OWL 2 Web Ontology Language 2012), including richer data types and data ranges, qualified cardinality restrictions,

K. D. Bimson (✉) · R. D. Hull
Intelligent Software Solutions, Inc., 5450 Tech Center Drive,
Suite 400, Colorado Springs, CO 80919, USA
e-mail: kent.bimson@issinc.com

K. D. Bimson
Department of Electrical & Computer Engineering, The University of New Mexico,
Albuquerque, NM, 87131-0001, USA

R. D. Hull
e-mail: richard.hull@issinc.com

M. Workman (ed.), *Semantic Web*, DOI 10.1007/978-3-319-16658-2_6

asymmetric, reflexive, and disjoint properties and enhanced annotation capabilities, increase expressivity but require additional practitioner investment.

The purpose of this chapter is to characterize some of the significant gaps between NL and ontology semantics, and to articulate some of the challenges that these gaps present to the representation of knowledge extracted from text sources. It should be noted that we are not talking about how well we can or cannot do "text analysis," rather how well we can represent NL meaning as RDF or OWL knowledge representations.

The motivation for clearly defining these gaps is based on hard lessons learned in delivering semantic solutions to customers. These semantic gaps often represent major obstacles to meeting customer and user expectations, because the gaps are poorly understood, poorly communicated, or improperly addressed. If we can clarify these semantic challenges, we have a greater probability of agreeing on project requirements, functional expectations, and the next generation of research required to fill the gaps. A few customer-related problems are summarized here to provide a context for the more technical discussion below.

Challenge 1: Unrealistic Expectations
Customers, in our experience, do not understand how much "meaning" can be extracted from text sources using standard, ontology-based text processing, and ontology representation techniques. This can lead to disappointment, confusion, and, potentially, project failure.

Challenge 2: Confusing Terminology
The meaning of the term "semantics" is very different in NL (linguistics) and in ontologies (knowledge representation). Our customers—as well as the modelers themselves—have a difficult time distinguishing between the two definitions, a confusion that also leads to unrealistic expectations. We recommend specializing the term "semantics" into "NL (or linguistic) semantics" and "ontology semantics." The arguments for this differentiation are the subject of this chapter. As semantic technology professionals, we should be careful about defining these terms more meaningfully for customers, users, and ourselves.

Challenge 3: Semantic Modeling, Mapping, and Knowledge Extraction Shortfalls
The semantic expressiveness of ontologies simply is not sufficient to represent the semantic complexity of NL, at least not without building significant "representational scaffolding" to support it, leading to severe language-to-ontology mapping and modeling challenges. These challenges lead, in turn, to problems in extracting knowledge from text sources and representing it as ontology constructs.

To borrow an analogy from the film industry, editing NL semantics enough to fit into standard ontology structures requires us to leave a significant amount of valuable knowledge on the editing room floor. Understanding the semantic trade-offs that must be made is critical for customers, information architects, and users, because meaning will be lost in the process of transforming NL semantics into ontology semantics, meaning that is often important to stakeholders. The remainder of this chapter outlines some of the major differences between "NL semantics" and "ontology semantics." A good starting point for this comparative analysis is defining the different levels of semantic representation in an NL.

6.2 Levels of NL Semantics

In order to compare NL semantics to ontology semantics, it is important to define the primary linguistic structures that carry meaning. More specifically, we define NL semantics as the meaning expressed within NL at the morphological, lexical, syntactic, and discourse levels. We begin with a simple definition of "NL semantics," although a full definition, explanation, and defense of this definition is not possible within the scope of this overview. Following is a brief summary of the first three levels at which meaning is expressed in NL:

- Morphological level: morphology refers to the internal structure of words in an NL, consisting of their component "morphemes," the smallest meaning-bearing units in language. For example, the word "vehicles" consists of two morphemes: the root morpheme "vehicle" and the plural suffix "s."
- Lexical level: a language's vocabulary, including words and, perhaps, fixed expressions. Words have one or more morpheme, such as the two-morpheme word "vehicles."
- Syntactic level: rules for constructing meaningful, well-formed phrases, clauses, and sentences in a language, including permissible word order, such as "large air transport vehicles."

Meaning is expressed at all of these (and other) linguistic levels, sometimes conjointly, and in ways that are often quite difficult for ontologies to represent, at least in an ontology's standard form. Semantic translation problems at each level are summarized below.

6.3 Morphological Level

Definitions and Backgrounds

The two major classes of NL morphology are inflectional and derivational morphology. Inflectional morphemes change a word's grammatical category without changing its grammatical class. Examples from English include noun plurals (truck > trucks), verb tense (work > work*ed*), and verb aspect (go > go*ing*). In each case, if the uninflected form is a noun (e.g., cat), the inflected form remains one as well (e.g., cats); if it is a verb, it remains a verb in the inflected form, and so on.

By contrast, derivational morphemes change the grammatical class of a root word, say from a noun to a verb. Examples from English include changing from a verb to a noun (derive > deriva*tion*), from a noun to an adjective (derivation

> derivation*al*), or from an adjective to an adverb (derivational > derivational*ly*). English has a highly productive morphology, which allows speakers to easily create new words and meanings simply by using morphological rules of composition. The meaning represented in NL morphology poses many challenges in translation to ontology representations, discussed next.

Ontology Challenges: Morphology

The reason that morphological semantics causes semantic challenges for ontologies is that *ontologies do not have morphologies*. The myriad meanings represented via morphology in NL, therefore, are quite difficult to represent using standard ontology constructs. For example, a class (such as vehicle) is neither singular nor plural in an ontology, and a relation (such as drives) cannot be inflected for tense or aspect, as in *drove* or *driving*. The result is that these morphological meanings in NL (i.e., number, tense, aspect, etc.) are not easily translated into ontology constructs, unless extra work is done to extend standard ontology constructs to do so. A relation, such as: *manufactures* in the RDF triple::*person:manufactures:equipment*, is neither present, past nor future tense. Nor does it represent a completed, ongoing, or habitual activity. There is simply no natural way to "inflect" the relation for tense and aspect, as we do for NL verbs.

The point is that representing the meaning inherent in NL morphology is *unnatural* for an ontology at best, and therefore it is not normally represented at all. The inability to represent an NL's many inflected meanings in an ontology results in a many-to-one semantic mapping challenge in translating from an NL to an ontology, as shown in Fig. 6.1.

One might argue that the sentence patterns in Fig. 6.1 differ also by auxiliary verbs, such as *will, have*, and *be*. Two points counter this objection: (1) ontologies do not naturally handle auxiliary verbs, either, and (2) many languages express these meanings via inflectional morphemes, even if English does not. In either case, the point is the same: Multiple specific NL meanings get compressed into one generic canonical form in the ontology.

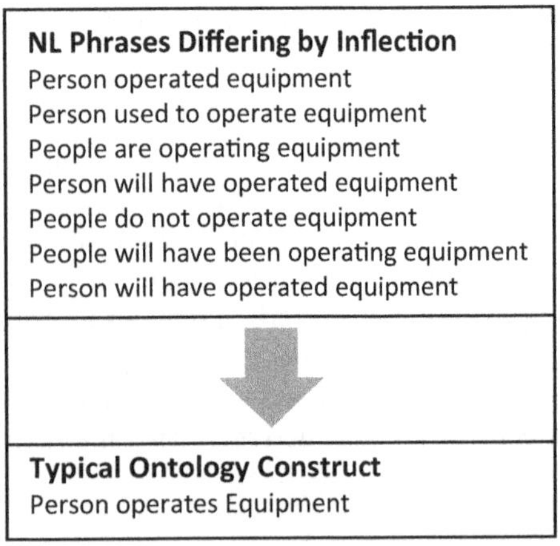

NL Phrases Differing by Inflection
Person operated equipment
Person used to operate equipment
People are operating equipment
Person will have operated equipment
People do not operate equipment
People will have been operating equipment
Person will have operated equipment

Typical Ontology Construct
Person operates Equipment

Fig. 6.1 Many inflectionally related NL words and meanings get compressed into one ontology construct and meaning, eliminating variations in meaning expressed by NL inflections, a process we call "morphological conflation." *NL* natural language

The point of this section is that significant meaning is expressed in NL quite naturally through its morphology, or closely related auxiliary verbs. Through morphemes, we can describe an activity as past, ongoing, habitually repeated, or a future improbability. We can identify subjects or objects that are individuals, couples, or groups. We can articulate whether an action is completed, ongoing, or not started at the time referenced. By adding modal verbs (such as *must* or *may*) and adverbs, we can discuss possibility, probability, certainty, or impossibility. And we can very flexibly combine these meanings (via morpho-syntactic rules) to express complex meanings like "The driver *most probably will have been operating* equipment by tomorrow morning." Without significant work, each of these NL grammatical patterns and meanings, when extracted from text, will be reduced to one canonical form, or assertion, in the knowledge base, such as *::person:operates:equipment*—leaving representation of the other morphological meanings (or senses) on the cutting room floor. For some applications, this may be quite acceptable, such as simply finding text sources that seem to be "about something" in general. In others, such as a global disease spread application, it would be advantageous to differentiate projected future state from reported past state, or the conditions under which the disease spread may be expected to increase or decrease.

A more extensive set of "lost meaning" examples is presented in Fig. 6.2, which illustrates the semantic downsizing required to map the semantics of NL morphol-

NL Morphology	Examples	Meaning	Ontology Construct
Inflections			
Tense	was located	past	
	is located	present	
	will be located	future	
Tense & Aspect	was being located	past progressive	locate (relation)
	is being located	present progressive	
	will be being located	future progressive	
	will have been being located	future past progressive	
	(and so forth)		
Number	location	singular	location (class)
	locations	plural	
Gender	he	masculine PN	
	she	feminine PN	
	they	non-gendered plural PN	*x no pronouns;*
Person	I	first person	substitute instance
	you	second person	name
	he	third person	
Possession	his	possessive PN	
	person's	possessive noun	Anton posseses X
	Anton's	possessive personal N	(assertion/relation)
Derivations			
	locate	verb	
	location	noun	location (class)
	locational	adjective	locate (relation)
	locationally	adverb	

Fig. 6.2 Many NL morphological variations are conflated to a single structure and meaning in an ontology. *NL* natural language, *SVO* subject–verb–object

Natural Language	Conflation Type	Ontology	Description
Morphological variants: ➢ Send ➢ Sent ➢ Sends ➢ Sending ➢ Manufacturer ➢ Manufacturers	Morphological conflation	Single canonical form ➢ Sends ➢ Manufacturer	Many inflectionally related forms in a NL are conflated to one morphological form in an ontology.
Phase of words: Steel manufacturer sends steel products to construction customers.	Lexical conflation	Single string literal forms: Class: steel-manufacturer Property: sends-steel-products-to Class: construction-customer	Multiple NL words are conflated to one ontology string for each literal (S, V and O)
Complex sentence syntax: Steel manufacturer sends steel products to construction customers.	Syntactic conflation	Simple SVO syntax: S: Steel-manufacturer V: sends-steel-products-to O:construction-customers	Complex NL phrase structures are conflated to simple SVO ontology statements (triples).

Morphological conflation cascades to lexical level

Lexical conflation cascades to syntax level

Fig. 6.3 Cascading conflation is produced by the effects of morphological conflation on the lexicon, and the effects of morphological and lexical conflation on the syntax of an ontology

ogy into simpler ontology constructs. Although this example set is clearly incomplete, it illustrates the many meanings in NL morphology that must be compressed into the simple semantics of an ontology, some of which are briefly discussed below. In addition, this corpus illustrates the many different word forms that are downsized to one canonical form in the translation from NL to an ontology.

We call this downsizing process "conflation," which has a dictionary definition of "a merging of diverse, distinct, or separate elements into a unified whole" (Dictionary.com 2014). In these examples, the various senses represented by morphological variations are conflated to one "ontological sense," and the word structures expressed by these morphological (and auxiliary verb) variants are conflated into one "ontological form." In other words, an ontology modeler will usually represent "go, goes, gone, going, went, will be going, will have gone, will have been going" as a single form with a single sense, such as "goes." To differentiate between sense and form conflation in discussing their impacts on ontology modeling, we use the phrases "semantic conflation" and "structure conflation." However, both are the result of the same root cause: The fact that an ontology has no morphology.

As we shall see, conflation at the morphological level has a cascading effect into the lexical and syntactic levels, since these linguistic levels are related in sophisticated ways. The end result is "cascading conflation" into the lexical and syntactic levels of ontology representation (Fig. 6.3), which we explore in the following sections. But, first, we discuss a few more semantic modeling challenges at the morphological level.

Related Word Challenge

Words related by morphological rules in NL must be represented as separate, unrelated constructs in an ontology, if they are represented at all. There is no principled way to "derive" a new concept by adding a meaningful morpheme, such as by adding an [-er] suffix to the verb *manufacture* to derive the noun *manufacturer*. If these two concepts are represented in ontology, they are represented as independent, unrelated terms, in the morphological sense. The ontology simply does not know that these terms share a common root meaning.

At best, their relationship could be expressed in that the class manufacturer might be modeled as the domain for the relation *manufactures*, as *in::manufactures rdfs:domain:manufacturer*, but this assertion states that *manufacturer* can be the "subject" of the relation *manufactures*, not that these two words share a common meaning. The shared "sense" between the two is simply lost in the ontology. In order to simplify presentation in the remainder of this chapter, we use a pseudo-code shorthand notation for expressing RDF triples, as in: *manufacturer manufactures product*, rather than using the standard RDF syntax::*manufactures rdfs:domain:manufacturer* and:*manufactures rdfs:range:product.*

Tense and Aspect Challenges

While we may make existential statements in an ontology, such as: *person operates equipment*, we are severely limited in representing whether this assertion is past, present, or future tense, as in "Rich operated the forklift" (past) or "Rich is operating the forklift" (present progressive). The meaning associated with such inflections is hard to represent in typical ontologies. In order to do so, temporal semantics must be added to the ontology (Hobbs and Pan 2006), along with the logic needed for inferencing about time, based on time–date property values. Although we can time-stamp a fact in RDF, such as: *event occurs-on date*, we must apply axioms to determine whether that event is a historical or future fact relative to any other assertion in the knowledge base. Tense is not carried in the relation itself, as it is in the NL (English) verb.

Ontology Modeling Challenges

Modelers must select class and property names from among the many possible inflected forms in naming ontology literal constructs. The following two triples are typical examples:

- Ontology 1: *Person operates equipment*
- Ontology 2: *People operate equipment*

In both cases, modelers intend to represent *person* and *equipment* classes and an *operate* relation that semantically connects them. Therefore, from a conceptual viewpoint, these two statements are semantically equivalent, a fact that is evident only to morphologically competent humans. In an ontology, assertions must be made—or algorithms must be developed—to identify this semantic equivalence; the semantic mapping is unnatural to the native ontology representation. For example, one might use the *equivalentClass* relation to indicate person and people are "synonyms," as in: *Person equivalentClass People*. However, this in no way represents the meaning inherent to the NL plural morpheme; in an NL, these are not equivalent words, as they become in the ontology, another example of morphological conflation. Ontologies simply do not have a natural way to represent morphological meaning variations of this kind.

Morphology Challenge Summary

Morphological conflation results in a significant meaning loss between an NL and ontology. In addition, it has a cascading effect at the lexical and syntactic levels, which we shall discuss momentarily. This meaning loss has significant ramifications for customer expectations. Whether in financial market analysis or military intelligence analysis, there is a significant difference between a fact in the past, present, or future tense. Knowing a stock merger will happen, or that an attack may occur, are very different than stating that they already happened, or that they probably will not occur. Yet, to represent these important meaning differences in an ontology requires a significant amount of additional "representational scaffolding," such as temporal extensions to ontology standards (Hobbs and Pan 2006), assertion of additional facts, or the addition of logic to handle comparative inferences among assertions, and so forth.

We discuss the approaches we are taking to extend an ontology to model morphological semantics in Chap. 7.

6.4 Lexical Level

Background and Definitions

There are significant semantic gaps between an NL and ontology at the lexical (or word) level as well. The morphological gaps between an NL and ontology play a significant role in that difference, as mentioned earlier, since people use morphological rules to create related words. These are therefore interrelated conflation challenges. An NL's lexicon, or vocabulary, can be roughly categorized into two primary types of words: content words and function words. Content words are those with lexical meaning (such as *car, cake,* or *careful*) while function words are those that relate one grammatical structure or meaning to another in some way (such as *and, then, will, therefore, he* or *of*). Although function words have no content, they do provide contextual meaning, expressing notions like conjunction, exception, direction, definiteness, or previous reference.

The set of function words is a "closed set," in that there are a set number of pronouns, prepositions, or articles in an NL. Whereas content words contain most of the domain meaning in an NL at the lexical level, function words are critical to how people understand the complex, meaningful relationships among content words, phrases, and clauses. Function words allow NL speakers to create increasingly complex, but meaningful, phrase and sentence structures, much as morphology does within words.

The problem is that an ontology has no function words, just as it has no morphology. It has no conjunctions to join two words, phrases, or clauses together to form conjoined subjects and objects or compound sentences, at least at the instance level of ontology representation. In addition, an ontology has no pronouns to support anaphora, or previous reference. It has no prepositions to turn nouns into modifiers of location, direction, or time. And it has no helping verbs to vary the tense, aspect, or modality of a verb phrase, as previously discussed. In summary, the lack of function words in an ontology makes it unnatural to represent these kinds of connective, directional, and referential semantics natural to languages, as exemplified by the italicized words in the following sentence: Tom *and* Alice drove *to* town *and then they* walked *to the* mall *to* buy clothes *and* see *a* movie *before* returning home *to* eat dinner, *after which they* went *for a* walk.

In this one example sentence, there are over ten kinds of "connective semantics" represented by function words like *and, to, then, they, which,* and *before,* as well as the phrases that these function words introduce, any one of which is difficult for RDF/OWL representations, especially in the A-Box (e.g., at the instance level).

The lexicon of an ontology is quite different from that of an NL in another way. Whereas NL words are meaningful to people, ontology words are not really meaningful to a computer, not in any linguistic sense (Manola et al. 2014). Ontology words consist of the strings that make up their class, property, and instance names. The fact that these strings "look" like NL words or phrases to people results from the fact that modelers usually choose to use strings based on NL words and phrases to help other people understand what sense they mean to express with a specific string. Within the ontology, however, a class named "equipment" could just as well be represented as the string "X" as far as the computer, RDF, or OWL are concerned, as long as it is unique within its own namespace (Manola et al. 2014). This leads to widely varying string names within different, but related, ontologies. Modelers are free to name a class or property whatever they like, as long as it is unique.

A third difference between a NL and ontology lexicon is that an ontology's words are not related by morphological rule, as NL words are in word sets such as *begin*, *beginner*, and *beginning*. Here we see how the lack of morphology has a significant representational impact on an ontology's lexicon. Each ontology "word" is semantically independent from all other words. This stands to reason: since an ontology has no morphology, there is no natural way to "create" new, related words via morphological rule, nor to represent the meaningful relationships among these inflected and derived word forms in the lexicon, as is done in an NL dictionary. Morphologically related words with a common core meaning—like create, creator, creative, and creation—have no more meaning in common in an ontology's dictionary than do words like disease, justice, drink, and sunrise, at least morphologically speaking. While one could imagine expressing morphologically related ontology constructs with RDF (pseudo-code) statements, such as:

- Creation noun-derived-from create
- Created past-tense-of create
- Creator creates creation

not only are these awkward and unnatural assertions but they also create other knowledge representation problems. For example, the first of these assertions uses the property *create* as the object of a statement, treating it as an individual and therefore pushing the overall OWL ontology into the OWL Full language, hindering description logics (DL) reasoning (Bechhofer et al. 2004). In addition, since there are no morphological rules of word composition, every morphological variation among words would need to be expressed as an individual fact, such as:

- Wanted past-tense-of want
- Created past-tense-of create
- Creating progressive-aspect-of create

This leads to an exploding knowledge base, if nothing else. As a final remark, it is hard to imagine how *creating* would then be used within the ontology, since the modeler wants to treat it as a present progressive verb form (a relation in the ontology), yet it has just been rendered a class within the ontology, which functions more as a noun form. This is a difficult challenge to address within the natural constructs of an ontology.

A final difference between NL and ontology lexicons is that an ontology's "words" often consist of what would be an entire phrase in an NL, such as the class name *major-launch-processing-operation* or the object property (relation) name *is-the-subject-of*. This string concatenation results from the fact that an ontology limits each class or property name to a single string, or literal, rather than a sequence of words making up a phrase. This "word concatenation" phenomenon represents the next level of cascading conflation—*lexical conflation*—which is discussed further in the challenges below.

Synonym Challenge

The three example assertions just presented point out another related semantic challenge for ontologies at the lexical level: Synonym identification and representation. Synonyms are two words with (roughly) the same meaning, at least within a specific communication context. Synonymy, natural to languages, is quite unnatural in an ontology. In this example, NL speakers may interpret *manufacturer, company,* and *organization* as roughly synonymous nouns and *develops, manufactures,* and *produces* as roughly synonymous verbs, given the context. Ontologies, however, do not naturally account for synonymy without an explicit assertion, such as (in pseudo-code): *Manufacturer same-as company*.

This representation is at best an inelegant and inefficient way to represent synonymy, but at least it is available within the representation standard. However, the challenge is greater for ontology class/property names that are based on concatenating words from NL phrases. Examples abound, such as the class name *Action-Temporal-Association* and the relation name *is-acted-upon-as-specified-by*, both from the US Department of Defense's JC3IEDM standard data model (Morris 2012). These are examples of lexical conflation, in which multiple NL word forms with multiple related senses are conflated to a single ontology form and sense. Since ontology constructs like these are not composed of individual, meaningful words, it is very difficult to map their meaning to NL paraphrases in text (such as the paraphrase: *Event relationship to time*). To do so would require a word-by-word, sense-by-sense comparison at the lexical level. To put it another way, ontology names are not "lexically based" but rather "string based." This example class name is not a combination of the three NL words *action, temporal,* and *association,* each with its own entry and meaning in the lexicon, and each belonging to a specific grammatical category (e.g., noun, adjective). Rather, the NL words were simply used by human modelers to form a string name, or literal, that looks as if it is formed from these words.

Lexical conflation, again, is caused by an ontology's lack of grammar, which is a set of rules that humans use to creatively construct phrases and sentences from individual words. Whereas in an NL, any number of words can work together as the subject of a sentence, for example, an ontology is limited to "one string," equivalent to one word in the ontology's lexicon. This limitation makes it very difficult to use conflated ontology concept names as a basis for finding NL paraphrases in text. The "one string–one meaning" restriction does not map well to "multiple words–multiple meanings" of an NL, at least without additional data structures and algorithms to support the analysis. Clearly, lexical conflation is a challenge that spans the lexical–syntactic boundary, and is discussed further in the section "Ontology Challenges: Syntax."

The implication of the synonym and paraphrase "gaps" between an NL and an ontology is that it is very difficult, using ontology semantics only, to identify, capture, and address synonymy and paraphrase in NL text. These gaps present major problems in extracting knowledge from text and transforming it into an ontology-based representation. Some of the knowledge extraction and representation problems that arise from this gap include:

- Difficulty in identifying and representing different words (from NL sources) with the same meanings (synonymy).
- Difficulty in identifying and representing different phrases with the same meanings (paraphrase) as lexically conflated ontology names.
- Difficulty in preventing redundant information extracted from NL text from being asserted to the knowledge base, since the redundancy is expressed by different words that mean the same thing in text.
- Difficulty in semantically integrating, or fusing, semantically related textual information, where the NL semantics cannot be naturally represented in the ontology.

Function Word Challenge in the Lexicon

The absence of function words in an ontology, such as articles (an, the), pronouns (he, she), or prepositions (of, from), makes it difficult to connect classes, properties, and triples in meaningful ways, particularly at the instance, or A-Box, level. It is difficult to express conjoined events and objects (and), conditionals (if this, then that), directionality (to the store), previous reference (the, he), or concurrency (while, during, when).

The fact that ontologies lack function words adds to the cascading conflation problem, contributing to some odd naming conventions in ontology modeling. Single NL phrases introduced by function words, such as prepositional phrases (PP), must often be split into two parts, with half of the phrase used in one construct's string name (e.g., a relation name) and half used in another's string name (e.g., a class name). Consider the following examples taken from the JC3IEDM standard (Morris 2012):

- Action is-the-subject-of Action-Functional-Association
- Action is-acted-upon-as-specified-by Organization-Action-Association
- Action is-geometrically-defined-through Action-Location

In the first example, the NL PP version of the relation (of action functional association) serves an adjectival role, modifying the predicate nominative subject. However, ontologies have no modifiers, such as adjectives, adverbs, or phrases that serve those roles. Within the ontology, the preposition "of" is therefore modeled as part of the relation name (*is-the-subject-of*), whereas the object of the preposition in the NL phrase is used to model the class name (*action-functional-association*). The NL PP has been split into two parts in the ontology, with one part used in the conflated relation name, and the other part used in the conflated class name.

Clearly, this "NL phrase splitting" is only part of the cascading conflation effect, since the predicate nominative (e.g., "the subject") is also conflated into the relation string name. This approach to naming ontology constructs results from the constraints placed on modelers at the ontology's syntactic level, a problem that is discussed further in the "Syntax-Level Challenges" section.

Lexical Challenge Summary

The lexicon of an NL and that of an ontology are significantly different in both form and meaning. One could claim that ontologies do not really express lexical meaning at all, at least not in any NL sense of the term. Even if we accept that they do, the depth and breadth of lexical meaning are highly restricted relative to NL for the reasons discussed. We present potential approaches to dealing with semantic gaps at the lexical level in Chap. 7.

6.5 Syntax-Level

Background and Definitions

An NL typically has a sophisticated syntax used to sequence words, phrases, and clauses to produce meaningful sentences. Syntax is governed by grammatical rules that define well-formed phrases and sentences for the language, allowing speakers to build words from morphemes; phrases and clauses from words; and sentences from phrases and clauses. Multiple words can be combined into a single grammatical construct, which can serve as a subject, predicate, or direct object of a sentence.

A subject in an NL is frequently a phrase with many words combined dynamically to express complex meaning; a subject in an ontology is always "one string." Humans understand that a combination of words in NL can be used together as a single subject or object (single referent), but that each of the constituent words has its own meaning in the lexicon as well, used within the phrase to specialize the meaning of the referent. That is, humans can parse these phrases into their meaningful words, and the words into their meaningful morphemes, based on shared rules of grammar. By so doing, they are also able to compare the meaning within these phrases to components of other phrases for similarities and differences. In that manner, people can easily determine that "minor launch processing operation" is similar to "major launch processing operation," perhaps different only in complexity or duration, based on the semantic difference between the words "major" and "minor."

Ontologies are quite different in this respect, as we have discussed. First, they have a very limited syntax. Second, they have no grammatical rules with which to construct phrases and clauses from individual words, just as they have no morphological rules for constructing new words from morphemes. These limitations have a significant impact on ontology modeling, as well as on mapping NL phrases and sentences to ontology constructs. Each is discussed in turn.

An ontology's syntax is typically limited to simple subject–verb–object (SVO) sentences (devoid of morphology), as in: *Mechanic repairs equipment.* These are called "triples" because they are assertions of exactly three ontology "words," each word a string, that are used to represent domain knowledge. Roughly, these are equivalent to simple active and passive voice sentences in an NL, with some significant limitations, however. We return in a moment to the challenges presented by this limited syntax.

Equally limiting is the fact that ontologies have no grammatical rules for dynamically creating new "multi-word" constructs, such as new class and property phrases, as discussed above. Therefore, ontologies cannot build phrases out of a sequence of words, and their associated meanings. For example, NL users can combine the words *lacrosse, sports,* and *equipment* into *lacrosse sports equipment,* and other speakers will immediately understand the phrase as composed of the constituent words and their individual meanings. In an ontology, *lacrosse-sports-equipment* is not composed of three words with three constituent meanings; it is one conflated string with one conflated meaning. Another class, say *sports,* is quite unrelated to it. The constituent word meanings are lost on the ontology, which has no grammatical rules to parse this string into constituent word structures and senses. Each string, to an ontology, is a "single canonical form" rather than a multi-word phrase, where each constituent word has a contributory meaning.

These two syntactic limitations result in syntactic conflation, in which myriad NL phrase and sentence structures must be transformed into simple SVO structures, with each S, V, and O represented by one (and only one) "ontology word." This is a serious restriction if our objective is to represent NL meaning within ontology semantic structures. A few of the impacts that these syntactic/semantic constraints have on ontology modeling include the following:

- Complex NL syntax must be transformed into a set of simple SVO sentences, or triples.
- Each construct in an SVO triple consists of one and only one named element, rather than an unlimited sequence of meaningful words combined into a phrase.
- Each construct in a triple is devoid of morphology, meaning that the S, V, and O do not vary in meaning, such as tense changes for verbs (relations) or singular/plural for subject and objects (classes).
- The lack of function words likewise makes it difficult to meaningfully connect one SVO triple to another to make compound sentences or to construct a discourse sequence, as in: Kent and Mark ate food. *Then* they played golf *before* they went to the movies.

In linguistic terms, this is like limiting NL sentences to noun–verb–noun structures, where the nouns and verbs can be at most one word in length and have no internal morphology to vary meaning. These combined limitations severely restrict the meaning that can be expressed naturally with traditional ontology syntax. The net result of cascading conflation at all of these linguistic levels is that modelers must fit some very large square pegs (NL semantics) into some very small round holes (ontology semantics), leading to significant meaning loss in the translation, as well as some rather bizarre modeling constructs.

Ontology Challenges: Syntax

Significant semantic content expressed by NL phrases and sentences will be lost in transforming them into an ontology's simple SVO syntactic structure, with morphological and lexical limitations adding significantly to the cascading conflation effects, as previously discussed. However, modelers use ontology conflations because they are attempting to meet two important but incompatible requirements: (1) to make an ontology name a single string representing a single class/property concept (a data structure modeling requirement) and (2) to make the string look like a normal, syntactically correct sequence of NL words that expresses the true mean-

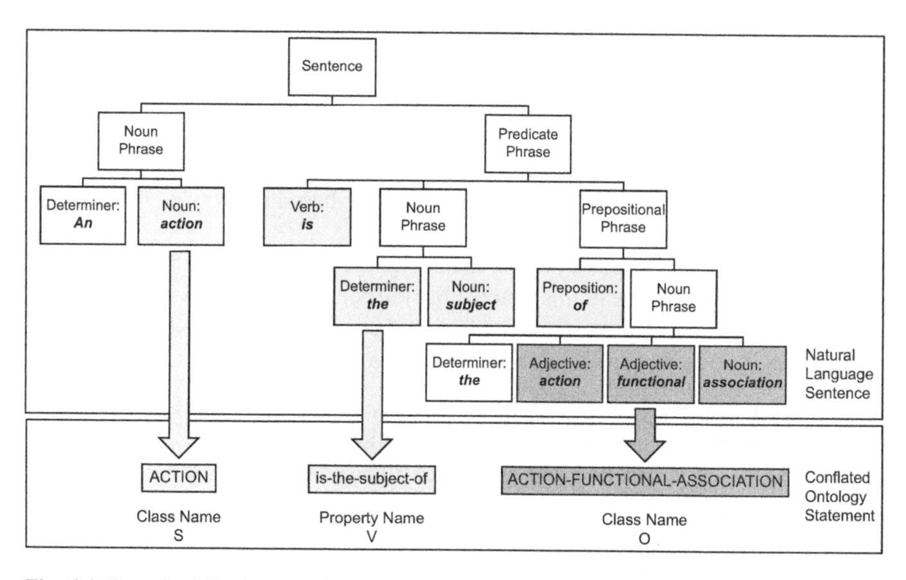

Fig. 6.4 Complex NL phrase and sentence structures. (Note: Complex NL phrase and sentence structures are conflated into three strings in an ontology, representing a simple SVO structure, with each element consisting of a single "ontology word.") *NL* natural language

ing of the string, so that people can understand the concept (a human understanding requirement). The resulting syntactic conflation, however, creates a serious confusion for those who do not understand that the ontology string names are not actually composed of individual NL words, as they appear to be.

Figure 6.4 illustrates syntactic conflation across all elements of a SVO triple, based on the first JC3IEDM example presented above, as it relates to the syntactic structure of the equivalent NL sentence. In effect, the NL syntax tree is flattened into three ontology strings, representing its SVO (or Class Relation Class) structure. The original NL sentence loses its internal constituent phrase structure. This results in the "phrase splitting" phenomenon discussed above, in which prepositions and predicate nominatives, for example, are conflated into the relation name, while the object of the preposition is conflated into the class name. The entire sentential structure has been conflated to three ontology "words" within this simple SVO triple.

6.6 Summary and Value Proposition

In summary, it is important to understand the significant differences between NL semantics and ontology semantics in order to level-set expectations for customers, users, modelers, and other stakeholders. Standard ontology constructs are too restrictive in structure and semantics to naturally represent the range of meaning that can be expressed in a NL, leading to cascading conflation problems in translating NL morphological, lexical, and syntactic meanings into ontology semantics. We have illustrated some of the types of "semantic gaps" that exist between an NL and ontology, and summarized some of the typical conflations resulting from those gaps. Given the complexity of mapping NL semantics into ontology semantics, readers may ask a legitimate question: Why extract ontology-based knowledge from text at all? Where is the value proposition?

A thorough answer to this question is beyond the scope of this short overview, and is not the main purpose of this chapter, but the benefits are significant and worth a summary comment. Here are only a few of the ways in which NL-to-ontology semantic translation can be scoped and extended to be highly useful:

- Extract only highly relevant essential elements of information (EEIs) from text. This might include specific event types, individuals, or locations, for example. This can be accomplished effectively for focused analysis needs by extending ontologies' domain-specific vocabularies and grammars, often available as open-source tools (Cunningham 2014).
- Relate extracted EEI's to each other based on the ontology. These connection graphs provide a way to semantically link, or fuse, EEIs extracted from heterogeneous sources based on shared concepts and relations.
- Extract and classify data based on a taxonomy, which provides more generalized search over text and data.
- Build rules for deductive reasoning over RDF knowledge bases, providing a way to infer new facts based on known facts.
- Use ontologies to publish the meaning of information for discovery by Semantic Web services.

In our follow-up chapter, we address some of the challenges that have been discussed in this chapter. In each case, our research objective is to bridge the gap between NL and ontology semantics, creating a more "natural" ontology representation of NL semantics.

References

Bechhofer, S., et al. (2004). OWL web ontology language reference, W3C recommendation. http://www.w3.org/TR/owl-ref/. Accessed 9 Nov 2014.

Bimson, K. D. (2009). Principles of interontology development, research supporting lingua franca requirements and design, tech. report, Modus Operandi.

Cunningham, H., et al. (2014). Developing language processing components with GATE version 8 (a user guide). University of Sheffield Natural Language Processing Group. https://gate.ac.uk/sale/tao/#sec:howto:plugins. Accessed 9 Nov 2014.

Dictionary.com. (2014). Dictionary.com. http://dictionary.reference.com/browse/conflation. Accessed 9 Nov 2014.

Hobbs, J. R., & Pan, F. (2006). Time ontology in OWL, W3C recommendation. http://www.w3.org/TR/owl-time/. Accessed 9 Nov 2014.

Manola, F., Miller, E., & McBride, B. (2014). RDF 1.1 primer, W3C working group note. http://www.w3.org/TR/2014/NOTE-rdf11-primer-20140624/.

Morris, M. (2012). Multilateral interoperability programme, the joint C3 information exchange data model (JC3IEDM). Greding, Germany.OWL 2 web ontology language, W3C recommendation. http://www.w3.org/TR/owl2-overview/. Accessed 9 Nov 2014.

RDF—Semantic Web Standards. (2004). RDF—semantic web standards W3C recommendation. http://www.w3.org/RDF/. Accessed 9 Nov 2014.

Chapter 7
The Lexical Bridge: A Methodology for Bridging the Semantic Gaps between a Natural Language and an Ontology

Kent D. Bimson, Richard D. Hull and Daniel Nieten

7.1 Introduction

Recently, a significant amount of research has been focused on extracting knowledge from natural language (NL) text and transforming it into an ontology-based semantic representation (Bimson 2012). The purpose of this research is to find ways to translate meaningful information in NL sources into a standardized, structured knowledge representation for the purposes of semantic normalization, integration, analysis, and reasoning.

However, a major obstacle to successfully translating NL meaning into ontology representations is that languages are semantically much more expressive than ontologies, resulting in significant meaning loss when translating NL semantics into ontology semantics. In Chapter 6 of this book, we characterize the kinds of meaning that get lost in the translation of NL semantics into ontology semantic structures, the reasons for that loss, and the impacts that these "semantic gaps" have on an ontology's representation of NL semantics.

The purpose of this chapter is to present a methodology that serves as a first step in spanning those semantic gaps, which we call "building a lexical bridge" (LB) between the NL and ontology representations of meaning. The goal of building the LB is to capture more of the meaning expressed in NL within an ontology. Our

K. D. Bimson (✉) · R. D. Hull
Intelligent Software Solutions, Inc., 5450 Tech Center Drive, Suite 400, Colorado Springs, CO 80919, USA
e-mail: kent.bimson@issinc.com

K. D. Bimson
Department of Electrical & Computer Engineering, The University of New Mexico, Albuquerque, NM, 87131-0001, USA

R. D. Hull
e-mail: richard.hull@issinc.com

D. Nieten
Red Lambda, Inc., 2180 FL-434, Longwood, FL 32779, USA
e-mail: dnieten@redlambda.com

M. Workman (ed.), *Semantic Web*, DOI 10.1007/978-3-319-16658-2_7

objective is to “lexicalize the ontology” by parsing ontology literals (i.e., class and property string names) into lexical items that can be used to generate an ontology lexicon (OL). The OL is used by our semantic equivalency algorithm (SEA) to compare the lexical meaning embedded in ontology literals to NL sources, in order to find synonymous and paraphrastic expressions in text. Together, the OL and SEA can be used for a number of high-value purposes, such as:

- Enriching the semantics of the ontology
- Improving semantic search of text sources based on the ontology
- Improving the results of ontology-based text extraction algorithms
- Enhancing our ability to compare the semantics of one ontology to that of another, and
- Identifying and eliminating redundant knowledge, such as synonymous Resource Description Framework (RDF) assertions

Each of these benefits is discussed in more detail in the *Potential Applications* section below.

7.2 Technical Approach: The LB

In this chapter, we present a method for *lexicalizing an ontology,* by which we mean *building a LB between an NL and an ontology*. The purpose of building a LB is to enhance the ability of ontologies to represent the lexical meaning hidden in class and property literals (or string names). By doing so, we prepare ontology constructs for word and phrase-based comparison with NLs.

Lexicalized Ontology Example

A simple example of “lexicalizing ontology constructs,” would be translating the ontology property name:[1]

```
records-observed-results-of¹
```

into its component NL lexical item stem forms:

```
[lexical item stems: record, observed, result, of]
```

and then marking them with their individual meanings, parts of speech and synonyms:

```
[lexical item: record
      (part of speech: transitive verb)
      (sense: set down in writing)
      (synonyms: write, pen, jot)]
```

A table-type representation of these constructs is illustrated in Fig. 7.1.

[1] Example from the approved *Joint Command, Control and Consultation Information Exchange Data Model (JC3IEDM)* triple [Action records-observed-results-of Action-Effect].

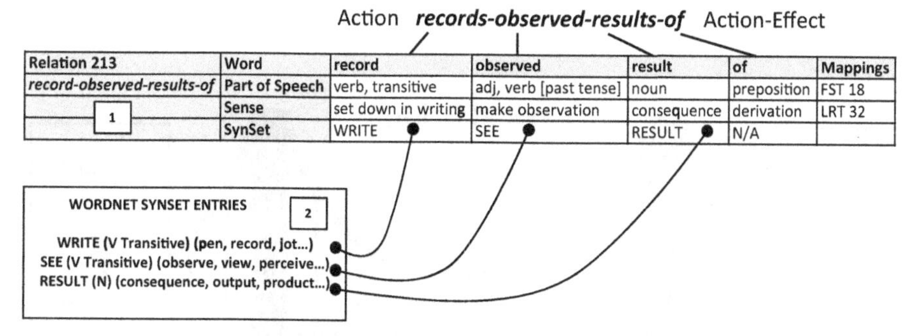

Fig. 7.1 The first step building a semantic bridge between the meaning expressed in a NL and an ontology is to lexicalize the ontology, parsing its class and property strings into constituent NL words, and then mapping the ontology lexicon to text. *NL* natural language

In order to build a LB, we need a way to transform ontology class and property literals (strings) into the NL words that modelers used to define these literals. Although NL-sounding string names are not needed to form unique Web Ontology Language (OWL) or RDF literals[2], the fact is that NL words and phrases are usually used by human modelers as the basis for these strings in order to make ontology class and property names more understandable to humans. This modeling approach can be exploited to our advantage in bridging the semantic gaps between an ontology and a NL.

Once the ontology string is lexicalized, each word extracted from the string can then be mapped to its sense (or meaning) and to its synonyms, using a thesaurus-style application—such as WordNet or FrameNet—each of which provides a database of English words and their synonym sets (also called synsets).

LB Components

A number of components are needed in developing the LB. These components, illustrated in Fig. 7.2, include:

1. *An ontology string parser*: The algorithms needed to parse ontology literals, or string names, into NL words in order to populate the OL.[3]
2. *A NL parser*: The algorithms needed to parse NL text into words, parts of speech, and senses (meanings).

[2] Literals can be any string in RDF and OWL, so *records-observed-results-of* could be represented as the string "x," as long as it is unique, though this is relatively useless for humans.

[3] As will be discussed later, this is not as straightforward as simply applying a NL parser to ontology strings, since the strings are often formatted differently (hyphenation in our example) and are often ungrammatical, in a NL sense, which leads to unsuccessful parsing results when using standard NL parsers.

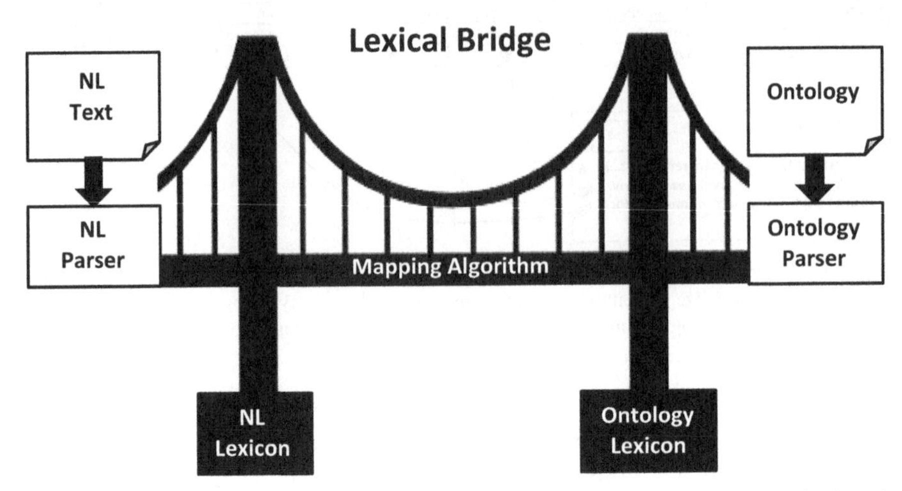

Fig. 7.2 The LB provides a semantic crosswalk from NL semantics to ontology semantics based on parsing ontology literals into NL words and creating an ontology lexicon. *LB* Lexical Bridge, *NL* natural language

3. *An ontology lexicon*: A lexicon of the NL words used to construct ontology literals, along with their intended parts of speech and ontology senses.
4. *A NL lexicon*: An online NL dictionary and thesaurus, providing access to synonyms for the OL.
5. *An ontology-to-NL mapping algorithm*: The algorithm for comparing words (and meanings) in ontology strings to NL words and phrases with similar meanings.

Once NL words (and their meanings, or "senses") have been extracted from ontology strings—and archived in the OL—they can be used as a basis for mapping the semantics of the ontology to the semantics of NL on a word-for-word basis. Our approach to this process is discussed below.

Building the LB

The first target application for our LB is to translate RDF triple class and property literals into NL words and senses. This application was developed on a project sponsored by the US Navy SPAWAR—called the RDF Find, Filter and Format (RDF-F3) project—under the Navy's Small Business Innovation Research (SBIR) Program.

The goal of RDF-F3 is to prevent redundant RDF triples—when extracted from text—from being asserted to the knowledge base. In order to do so, we must be able to map the meaning of RDF literals in the knowledge base to the meaning in text from which new assertions will be extracted. The LB will accomplish this goal by translating RDF triple literals into NL words and phrases, allowing the meaning of the words embedded in RDF literals to be semantically compared to words from NL text. The technical objectives of the project were to:

1. Define RDF redundancy in a formal, semantic way
2. Develop the LB methodology
3. Design the LB architecture
4. Develop an LB prototype

Our accomplishments in each of these areas are discussed in the following sections.

Potential Applications of the LB

A lexicalized ontology provides a technical basis for significant improvements in ontology-based knowledge extraction from text sources. These improvements include, but are not limited to, the following potential applications, each of which contributes significant benefits to deployed semantic solutions.

1. *RDF redundancy identification and prevention.* The assertion of redundant RDF triples to a knowledge base creates excessive knowledge growth as well as disconnected graphs. The LB technology can be used to compare the meaning of RDF triple strings to NL words and phrases, identifying potentially redundant knowledge in text before it is asserted to the knowledge base. This use case is the primary objective of the RDF-F3 project and is discussed further below as the initial application of the LB.
2. *Improving semantic search in text.* The LB's NL words representation of RDF constructs can be used to find synonyms and paraphrases in text sources that are beyond the scope of current ontology-driven search engines.
3. *Learning new classes and properties.* By lexicalizing class and property names, we will be able to significantly improve the identification of semantically similar, but nevertheless different, phrases in text. These can be recommended as a "new ontology relation" or as a "new class" (as mentioned above). In other words, our architecture provides a solid foundation for "learning" new ontology constructs. This can be done much more effectively using the LB than with native RDF constructs, since it does word-based analysis and comparison.
4. *Bridging the semantic gap between ontology semantics and NL semantics.* Our research (Chapter 6) has shown that significant meaning is lost in transforming NL semantics into ontology structures. By adding a "lexicon," "thesaurus," and "paraphrase" data structures to ontologies, we provide a significant LB between the rich semantics of an NL and the simple semantics of an ontology.
5. *Improving service-based semantics.* By providing service-based access to an ontology's lexicalized structures, we expose the meaning within class and prop-

erty string names for exploitation by other applications and analysis algorithms, improving an external application's "understanding" of what the ontology really means.

6. *Semantic comparative analysis of ontologies*. It is often quite difficult to automatically compare the semantic content of one ontology to that of another ontology. This is because modelers use different words and phrases to create ontology class and property string names. By lexicalizing ontologies, we translate these string names into their component NL words and phrases, thereby improving inter-ontology comparative analysis based on NL semantics.

In the remainder of this chapter, we focus on applying our LB methodology and algorithms to RDF redundancy prevention as a first target application.

RDF Redundancy Definition

One of the keys to growing robust, lean, nonredundant knowledge bases is identifying text that is semantically equivalent with knowledge already in the triple store, as well as identifying new, ontology-relevant knowledge that should be asserted to the knowledge base. In other words, we must find ways to differentiate between redundant and nonredundant knowledge, using the ontology as a reference semantic data structure. The major challenge of this task is comparing the meaning of text words (TW) and phrases to the meaning of words embedded in ontology literals, the purpose for which the LB is being designed. For this reason, we are applying the LB methodology and technology to RDF redundancy prevention as a first target application.

The first step in addressing this challenge was to formally define RDF redundancy and to use that formal definition as a basis for developing the LB use cases, architecture, and prototype. For purposes of brevity, we only summarize the definition of RDF redundancy in this chapter. RDF redundancy must be defined along two axes: (1) graph equivalence, and (2) semantic equivalence. Standards, such as RDF Primer (2004) and RDF Semantic Web Standards (2004), focus on the former, wherein equivalent "meaning" is based on equivalent RDF graphs. Graph equivalence, however, does not fully address semantic equivalence, which is based on linguistic synonymy and paraphrase rather than intersecting nodes and edges. In lexicalizing the ontology, we provide a lexical basis for comparing the "intended lexical meaning" of ontology class and property names by parsing them into their constituent NL words. These words can then be compared for similar meaning, either within the same triple store (RDF to RDF), across different triple stores (RDF to RDF′) or between a triple store and an NL corpus (RDF to NL text). The meanings could be identical (same words) or equivalent (synonyms) or different. In the former case, the RDF is redundant. In the second case, the RDF may be redundant or it may add valuable information, as discussed above. In the last case, the semantics of each triple is different, representing new knowledge.

We define *semantic redundancy* in RDF triples on both a class and an instance level, as follows:

Formal Definitions of Redundancy:

Class Redundancy: An RDF triple T at the class level (T box), represented as <C1 Relation C2>, is redundant with respect to a specific knowledge base KB IFF there exists a triple T, in KB, represented as <C3 Relation' C4>, for which Relation is identical to Relation' , AND C1 is a synonym for C3, AND C2 is a synonym for C4, as defined by the KB lexicon, L.

Instance Redundancy: An RDF triple T at the individual level (A box), represented as <I1 Relation I2>, is redundant with respect to a specific knowledge base KB IFF there exists a triple T' in KB, represented as <I3 Relation' I4>, for which Relation is identical to Relation', AND I1 and I3 identify an equivalent real work referent, AND I2 and I4 identify an equivalent real world referent, as defined by the KB lexicon L.

These definitions add a NL (semantics-based) equivalence definition to World Wide Web Consortium's (W3C's) RDF (graph-based) definition, providing a lexical basis for identifying identical, equivalent, and (potentially) redundant RDF assertions in the knowledge base, as well as semantically equivalent NL statements. This semantic definition will improve our ability to identify and filter out NL equivalents before assertion of the RDF to the knowledge base. It also provides us with the formal foundation needed to develop the LB methodology, use cases and architecture.

LB Methodology Applied to RDF Redundancy Evaluation

Our methodology focuses on using the LB to prevent the assertion of redundant RDF triples to the knowledge base, particularly when RDF triples are extracted from NL text sources. The concept is to parse RDF ontology class and property names into NL lexical items, using the latter as a basis for comparing the meaning of the embedded words in RDF strings to words and phrases in text sources, identifying potential synonyms and paraphrases. A lexicalized RDF triple that means the same thing as an NL statement identifies that NL text as identical (redundant) or equivalent (potentially redundant) relative to the existing knowledge in the RDF triple store.

Our methodology, illustrated in Fig. 7.3, is a step-by-step process that the analyst will use for redundancy prevention. The methodology is based on both the graph-based and semantics-based definitions of RDF redundancy. In this process, we use the LB Parser to parse ontology literals, lexicalize them, and apply our lexico-semantic analysis to the parsed set of terms to determine lexical equivalency for the RDF triples. This information can then be used to determine redundancy with respect to the individual RDF and associated graphs.

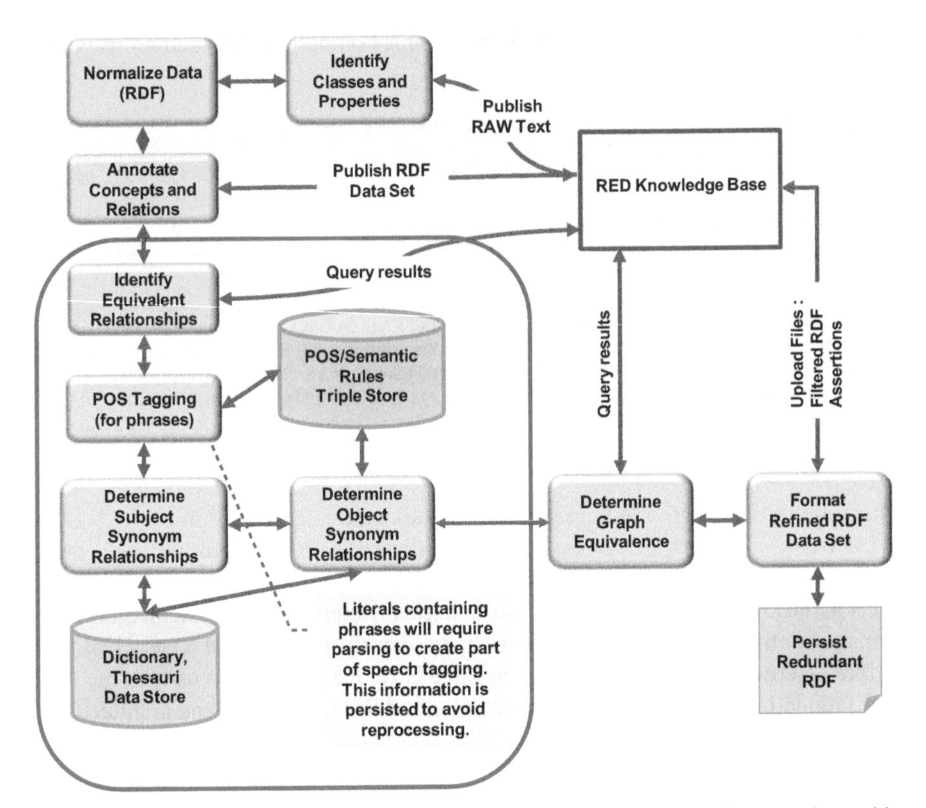

Fig. 7.3 LB methodology for lexicalizing RDF strings as a basis for semantic comparison with text sources with the goal of preventing the assertion of redundant RDF statements extracted from text. *LB* Lexical Bridge, *RDF* Resource Description Framework

The steps in our RDF redundancy analysis and prevention process are as follows.

Step 1 First, we apply the most straightforward criteria for determining RDF equivalency based on graph comparison, which involves leveraging OWL and RDFS constructs such as `owl:sameAs and owl:equivalentClass.`

Step 2 Second, we determine the lexical equivalence for class and property string names in one or more ontologies, beginning with the subject and object (classes) and ending with the verb (object property).

Step 3 Third, we determine the class or property literal equivalence, based on the W3C graph-based rules of equivalence.

Step 4 We then lexicalize literals and apply our rules of semantics-based equivalence as expressed in our lexical definition of RDF redundancy and our RDF equivalency algorithm, discussed below.

Methodology Data Products New phrase structure parse trees and rules are created during steps 2–4 and persisted, as are the subset of triples and associated graphs that have satisfied the equivalence criteria. At this point there are two artifacts of interest. The first is the filtered data set resulting from the removal of *identical* RDF (provably redundant) and the second is the collection of RDF and associated graphs that have been identified as *equivalent RDF* (potentially redundant), based on both graph-based and semantic-based algorithms. These artifacts provide additional data sets for analysis and potential human vetting to confirm or reject an RDF triple and its associated graph as redundant. The vetted and/or non-vetted data sets can be published back to the knowledge base for use by any other RDF analysis tools.

LB Architecture

The LB's conceptual architecture, illustrated in Fig. 7.4, leverages a number of components from a portfolio of text parsing, extraction, transformation, and analysis technologies used to develop ontology-driven NL solutions. Components labeled 2, 3 and 4 represent mature components. Those labeled 1 represent prototype components added to complete our LB and RDF-F3 prototype. This framework provides a mechanism for connecting processing resources across a message bus construct. The processing components are connected to the message bus, where each subscribes to one or more input topics and can publish processed results on one or more output topics. The processing components are focused on part-of-speech (POS) tagging for RDF literals, identifying the lexico-semantic information for each word and phrase in the literal.

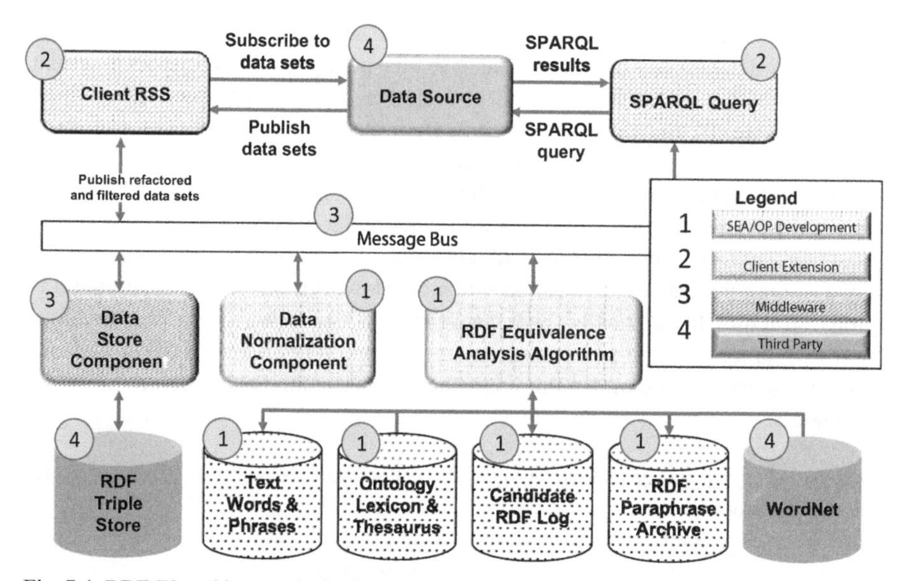

Fig. 7.4 RDF-F3 architecture lexicalized RDF triple literals for word-by-word comparison with text. *RDF* Resource Description Framework

LB Prototype and Results

Our initial prototype was designed to demonstrate three fundamental components of LB processing within the RDF-F3 application:

1. Ontology parser: Automated parsing of an RDF literal into its component lexical items
2. NL parser: Parsing semantically equivalent NL text into its component lexical items
3. Semantic mapping algorithm: Comparison of RDF lexical components to NL text lexical components to determine semantic equivalency.

As previously discussed, RDF class and property string names are not composed of words in the lexical sense. It is therefore difficult to compare a relation literal, such as JC3IEDM's relation `records-observed-results-of` (Multilateral Interoperability Programme (2009)), with words and phrases extracted from text, on a word-for-word basis, as illustrated in Table 7.1.

To do so, the literals must be parsed into individual words. In addition, candidate NL phrases must also be parsed into their individual words, as illustrated in the same table. The words from each must then be compared for equivalent meanings, or "senses."

Step 1: Parse Ontology Literal The first step in our prototype was to parse the string representing an RDF class or property literal, determining the constituent words and their parts of speech. Initial parsing of literals into words is illustrated in Table 7.2, rows 2–5 for the literal `records-observed-results-of`. We used an extended version of the Brill tagger (1992) to perform the parsing and the POS tagging, though we experimented with others as well.

Each word, together with its POS tag, was then processed to identify its potential meaning, which in WordNet means identifying the synonym sets (synset) to which it could belong. In this example, we show both the noun and verb synsets for the word `records`, as provided by WordNet.

It is important to note that none of the POS taggers performed correctly on this literal string. All identified `records` as a noun rather than a verb, very likely due to the fact that there were no other "subject nouns" in this literal, which parsers will treat as a grammatically well-formed sentence. However, this literal is an ungrammatical structure, linguistically speaking, at least in its ontology form. We therefore propose to tailor the Brill parser to account for "ontology literals" differently from "text sentences" to account for the linguistically ungrammatical structure of ontology literals.

Table 7.1 JC3IEDM object property string name and semantically equivalent NL used in the lexical bridge prototype

Ontology object property	Natural language paraphrases
`records-observed-results-of`	Wrote results about
	Will report observation concerning
	Are archiving observed evidence of

Table 7.2 Results of applying text parsing to ontology literals (records-observed-results-of) and to text (e.g., wrote results about). The text parser erroneously identifies "records" as the noun rather than the verb

Sentence/ ontology concept	Word #	Word	POS tag	POS	Root/lemma
Records observed results of	1	records	NNS	Noun, plural common	record
	2	observed	VBN	Verb, past participle	observe
	3	results	NNS	Noun, plural common	result
	4	of	IN	Preposition or subordinating conjunction	of
Wrote results about	1	wrote	VBD	Verb, past tense	write
	2	results	NNS	Noun, plural common	result
	3	about	IN	Preposition or subordinating conjunction	about
Will report observation concerning	1	will	MD	Modal verb	will
	2	report	VB	Verb, base form	report
	3	observation	NN	Noun, singu-lar common	observation
	4	concerning	VBG	Verb, present participle	concern
Are archiving observed evidence of	1	are	VBP	Verb, present tense, not 3rd person singular	be
	2	archiving	VBG	Verb, present participle	archive
	3	observed	VBN	Verb, past participle	observe
	4	evidence	NN	Noun, singu-lar common	evidence
	5	of	IN	Preposition or subordinating conjunction	of

Although not yet prototyped, our next objective for this lexicalization step to is to lexicalize an entire ontology in this manner, building a core Ontology Vocabulary by parsing class and property literals into NL words, together with their senses (or meanings). We will then add synonyms for each of the ontology's lexical items, creating an Ontology Thesaurus. The vocabulary and thesaurus, taken together, form the OL, as defined in our RDF redundancy definition.

Step 2: Parse Text Paraphrases The second step in our prototyping effort was to parse NL paraphrases for their lexical content, as illustrated for three paraphrases in Table 7.2, rows 5 to the end. Each of these words, together with its POS, was then used to identify potential senses, or synsets, in WordNet, as illustrated in Fig. 7.5 for both the noun and verb senses of "records."

Step 3: Compare Lexical Semantics in Ontology Literal to Lexical Semantics in Text The next step in our algorithm is to process the individual words in the ontology literal parse trees, comparing each ontology word (OW) to text words (TWs) in the text parse trees. Specifically, this involves comparing each OW's POS and synset to a candidate TW's POS and synset (Table 7.2 and Fig. 7.6). We call this the *SEA*. For this specific application, it is an *RDF SEA*.

Although we have developed the logical algorithm for RDF semantic equivalency analysis (discussed in the next section), it has not been prototyped because we have not yet developed a robust OL, which it needs to do the word-to-word comparative analysis.

Steps 2 and 3 beg the question of how paraphrase candidates are identified in text sources in the first place, since this needs to be an automated process. Our algorithm uses the OL as a filter to identify candidate synonymous/paraphrastic NL expressions after POS tagging and sense disambiguation have been performed. In other

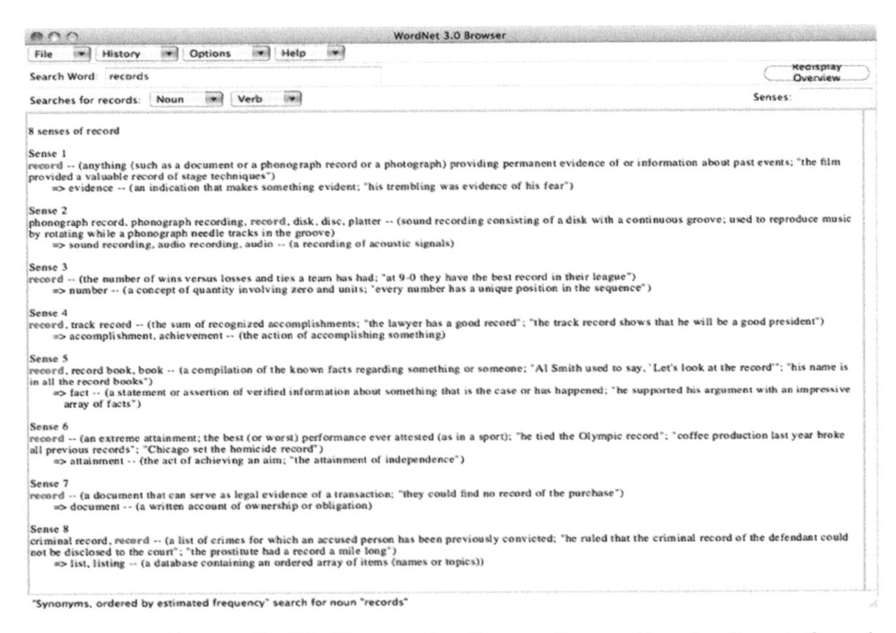

Fig. 7.5 A good lexicon, like WordNet, provides all senses for a word based on its part of speech

```
Algorithm: RDF-Equivalency algorithm
Function RDF-Equivalent (T: list of terms)
    for each {sentence || phrase ∈ T } do
        W := ParseSentence(Phrase)
        M := FindSynsets(W)
    end for
    for each { syn chain ∈ M } do
        P := SearchForSynsetChain(Phrase)
        s := BestSense(p,I,∅)
    end for

% Execute the Part of Speech tagging
Function ParseSentence (T: phrase)
    P := PartofSpeechTagging(T)

% Find WordNet Synsets - retrieves all the synsets associated with
the words
Function FindSynsets(W: list of word)
    M := 0
    for each {noun ∈ W || verb ∈ W || adj ∈ W || adv ∈ W} do
        s := Synsets(W)
        if s length > 1 then
            M := AddSynsets(M, FindSynsets(s))
        else
            M:= AddSynets(M,s)
        end if
    end for

% Determine the Best Synset chains
Function BestSense (t: term, I: list of senses, P: list of terms)
    BestSense := ∅ MinDistance := 0
    for each {sense s ∈ t} do
        d := MinDistanceS(s, I)
        if MinDistance = 0 or d < MinDistance then
            BestSense := s
        MinDistance := d end if
    end for

% Minimum distance using dijkstra's
Function MinDistance (s: sense, I: list of senses)
    d := 0
    for each {sense s' ∈ I} do
        d := d + DijkstraShortestPath(s, s')
    end for
```

Fig. 7.6 RDF equivalency algorithm uses the lexical bridge method to parse ontology literals into NL words and senses for comparison with text words and senses in order to identify potentially redundant triples in text sources. *RDF* Resource Description Framework, *NL* natural language

words, a TW from a text source is either in the OL or not, based on a look-up. This means that the TW is semantically equivalent to an OW in the lexicon. If the TW is not in the OL, then it is eliminated from consideration as a possible synonym. If the TW is in the OL, then it is retained for consideration as part of a synonymous (or paraphrastic) text expression. Once all candidate words are identified in the text source, N-grams will be used to determine whether individual words (like "wrote" in Table 7.2) occur in a *paraphrastic or synonymous context* with other OW's (such as "results" in Table 7.2), a process explained more fully in the next section about our SEA. This analysis involves iterative comparison of word senses among OW's and TW's. In this way, the OL will be used to filter out TW's with no meanings in common with OL words and to identify sequences of words from text that are synonymous with word sequences representing ontology literals (Fig. 7.5).

Semantic Equivalency Algorithm Our algorithm uses *synset chains,* or graphs of sense relationships, to compare the lexical semantics of ontology literals to the lexical semantics of text paraphrases. Since these chains are graphs, we can apply standard graph algorithms, such as Dijkstras (1959) and Hart (1968), to our semantic equivalency analysis (SEA). The SEA is responsible for assembling the synset chain for each noun, verb, adjective, and adverb contained in a literal (from the ontology), sentence or phrase (from text). The algorithm then searches the knowledge base to determine if the synset chain is already asserted. If the chain does not exist then the algorithm will assert the new chain and the associated RDF triple reference. If the synset chain does exist for one word in a given phrase, then the algorithm searches for each subsequent word's synset chain as well. An equivalent phrase will have the same synset chains. A version of the SEA tailored for RDF analysis, called the RDF-Equivalency Algorithm, is presented in Fig. 7.6.

7.3 Conclusion and Next Steps

In Chapter 6, we summarized the mismatches between natural language and ontology semantics. As a result of these mismatches, many kinds of NL meanings will be lost when attempting to represent them in standard ontology representations, severely limiting the effectiveness of ontology-based semantic search, knowledge extraction, knowledge representation, knowledge discovery and ontology-to-ontology comparative analysis.

In Chapter 7, we have proposed that the first step in overcoming these limitations is to build a Lexical Bridge between a NL and an ontology. The Lexical Bridge is composed of the lexical and phrasal data structures and algorithms needed to compare the words and phrases in NL text to those embedded in ontology literals. We demonstrated the use of the Lexical Bridge in determining semantic redundancy in an RDF triple store, with the goal of ensuring that ontology-based knowledge extracted from text was represented only once within the knowledge base, thereby limiting knowledge growth to uniquely different pieces of information.

Our early prototype demonstrated that this approach has promise. Though the Lexical Bridge has its limitations, such as its inherent complexity, it is an important step in providing a more structured ontology-based representation of NL meaning, which is an important goal in most aspects of semantic processing.

In summary, we propose that NL *words and phrases*, together with their basic *meanings*, be used to provide the basic building blocks for a semantic bridge between a NL and an ontology.

References

Bechhofer, S., et al. (2004). *OWL Web Ontology Language Reference*, World Wide Web Consortium (W3C) recommendation, Feb. 2004. http://www.w3.org/TR/owl-ref/.

Bimson, K. (2009) *Principles of interontology development, research supporting lingua franca requirements and design*, tech. report, Modus Operandi.

Bimson, K., Teresa N., Jon N., & David M. (2012). *Tactical Semantics: Extracting Situational Knowledge from Voice Transcripts using Ontology-Driven Text Analysis*, Semantic Technology Conference, San Francisco, CA, June 2012.

Brill, E. (1992). A simple rule-based part of speech tagger, HLT '91: Proceedings of the workshop on Speech and Natural Language, Morristown, NJ, USA: Association for Computational Linguistics, pp. 112–116, doi:10.3115/1075527.1075553, ISBN 1-55860-272-0.

Dictionary.com. (2011). www.dictionary.com. http://dictionary.reference.com/browse/conflation.

Dijkstra, E. W. (1959). A note on two problems in connexion with graphs. *Numerische Mathematik*, 1, 269–271. doi:10.1007/BF01386390.

Hart, P. E., Nilsson, N. J., Raphael, B. (1968). A Formal Basis for the Heuristic Determination of Minimum Cost Paths. *IEEE Transactions on Systems Science and Cybernetics* (SSC), *4*(2), 100–107. doi:10.1109/TSSC.1968.300136.

Lowe (1990). The Berkeley framenet project.

Miller, G. A. (Ed.) (1990). WordNet: An on-line lexical database. *International Journal of Lexicography 3*(4), 235–312.

Mulilateral Interoperability Programme. (2009). *The Joint C3 information exchange data model metamodel*. Greding, Germany.

RDF Primer. (2004). W3C recommendation, Feb. 2004. http://www.w3.org/TR/rdf-primer/#rdfmodel.

RDF Semantic Web Standards. (2004). W3C recommendation, Feb. 2004. http://www.w3.org/RDF/.

References

Chapter 8
Reliable Semantic Systems for Decision Making: Defining a Business Rules Ontology for a Survey Decision System

Pavani Akundi

8.1 Structure of Researched Data

Most people are familiar with three uses of survey techniques: the measurement of public opinion for newspaper and magazine articles, the measurement of political perceptions and opinions to help political candidates in elections, and market research designed to understand consumer preferences and interests (Fowler 1988). Survey modeling methods allow researchers to gather information and make informed decisions. A deeper qualitative analysis presents several other decision criteria used to formulate the question, including how the survey delivery methods impact the responder, and how the surveyor guides the survey. Inconsistency can exist with the target audience, the number of sensitive questions in the survey, survey length, and decision criteria.

Designing a good survey involves selecting the questions needed to meet and design research objectives (Fowler 1988). Qualitative, quantitative, and mixed methods are applied in generating and analyzing surveys. Quantitative methods provide the best evidence of statistical significance of decisions made through probability sampling, standardized measurements, and targeted surveys for special purposes (Fowler 1988). Qualitative methods are exploratory and offer an introspective view into how decisions are made through open questioning when the outcomes are not predictable (Mack et al. 2005). Focus groups and feedback surveys anticipating open comments are qualitative in nature, and reveal the behavioral aspects social survey researchers are seeking. Applying both methods is considered mixed survey modeling, and how the respondent chooses responds is the unknown variable.

At the most basic level, a survey consists of a series of questions and associated responses. The end goal is data collection by asking people questions and producing

---

P. Akundi (✉)
Department of Computer Science and Engineering, Southern Methodist University, Dallas, TX, USA
e-mail: pavakundi@yahoo.com

M. Workman (ed.), *Semantic Web*, DOI 10.1007/978-3-319-16658-2_8

statistics, representing the features of the study population (Fowler 1988). Survey data are categorized by response and nonresponse data, where the nonresponses can be indicative of poor wording, perceived misunderstanding, or perceived sensitivity. Social surveys offer researchers the ability to measure perceptions to understand how things are in a population. Special-purpose social surveys like health assessment surveys combine multiple variables like a target audience and sensitive questions to collect population statistics impacting social welfare. These surveys contain a greater opportunity to ask personal and private questions. Questions about income, unlawful behavior, sexual habits, substance abuse, and child rearing, for instance, are examples of sensitive questions. These questions may be perceived as sensitive questions leading to poor response rates due to the social stigmas associated with these behaviors. The following questions are considered perceived sensitive questions:

- What is your total annual household income?
- Has there ever been an investigation by child protective services (CPS) related to this child?
- Do you think this child has ever been physically abused?
- Do you think this child has ever been psychologically abused or mistreated?

The unknown variable in a social survey design is called a social desirability bias and records under and over-reporting in responses based on the survey administration technique. Scholarly research reveals individuals are more likely to falsely represent themselves in self-administered questionnaires, where perceived sensitive questions exist; however, findings show that people tend to be more accurate about revealing socially undesirable behavior in web forms, such as computer and voice-assisted surveys versus questionnaires. Socially undesirable behavior is partly a function of perceived question sensitivity, but question sensitivity depends on the respondent's actual status regarding the variable in question (Kreuter et al. 2008). The Community-Wide Children's Health Assessment and Planning Survey (CCHAPS) presented in this chapter builds on this research by analyzing the format of sensitive questions in a self-administered paper questionnaire.

The greater goal behind survey design in the nonprofit sector is collective impact. Collective impact requires large-scale organizations working together around a common agenda to solve complex social problems (Kramer 2011). This means working collectively to mutually reinforce activities rather than focusing on independent action as the primary vehicle for social change (Kramer 2011). Nonprofits are the voice of the people, where accessibility to improve social welfare is typically the primary objective. Surveying methods allow nonprofits to both participate and present concerns and issues in the community by obtaining statistically representative data from specific social groups.

The CCHAPS assessment team surveyed community leaders, parents of children aged 0–14 years, and focus groups with both parents and children to better understand the issues and concerns surrounding children's health. They split the survey into two versions to reduce the stress on the respondents to complete it in a reasonable time frame. Surveys were administered by mail, phone, and the Internet.

Only two surveys have been administered by the assessment team since 2008, and a third is due to be administered between August 2014 and April 2015 blending highly answered as well as unanswered questions. Several questions are selected from the national, Texas, and CCHAPS question bank. Although these versions contain question more redundancy than variability, the CCHAPS team expressed maintaining a majority of the same questions across three surveys reveals more patterns about the survey population versus two. In this chapter, we review the weight of the question format in relation to the responses to predict business rules. The variance between surveys is valuable only if additional questions were added to derive better responses.

We review the questions in greater detail to establish the business rules affecting all three surveys. The model depends on a baseline decision framework validating the current response data. Once the decision criteria and business rules are defined, we can begin to look for patterns in the content to shape future questions. Figure 8.1 illustrates the questions from 2008 explaining how children's health stacks up against the national and state levels. The question regarding hearing loss was modified in the CCHAPS survey from the national question: Does child have hearing problem? We see in all three surveys, there is zero variability in the response results although the question was modified. Therefore, we can conclude that this question is asked correctly and requires no change to initiate a stronger response. Equally, a marked difference between results requires further analysis to determine whether the results were based on true population statistics, poor wording,

CCHAPS National and State Comparisons of Children's Health	CCHAPS Results	National Results	Texas Results
In general, how would you describe your child's health?			
Excellent/very good	85%	84%	78%
Good	12%	12%	16%
Fair/poor	3%	4%	6%
Has a doctor or health professional ever told that your child has asthma?			
Yes, child has had condition at some point	18%	14%	11%
No, child has never had condition	82%	86%	89%
Has a doctor or health professional ever told that your child has bone, joint or muscle problems?			
Yes, child has had condition at some point	4%	3%	2%
No, child has never had condition	96%	97%	98%
Has a doctor or health professional ever told that your child has hearing loss?			
Yes, child has had condition at some point	3%	3%	3%
No, child has never had condition	97%	97%	97%
Has a doctor or health professional ever told that your child has blindness or other vision problems?			
Yes, child has had condition at some point	2%	2%	2%
No, child has never had condition	98%	98%	98%
Has a doctor or health professional ever told that your child has diabetes?			
Yes, child has had condition at some point	1%	1%	Not available
No, child has never had condition	99%	99%	Not available
How many days of school has this child missed during the past year because of health problems?			
None	36%	24%	30%
1-5 days	48%	58%	55%
6-10 days	8%	12%	10%
11+ days	8%	6%	5%

Fig. 8.1 Community-Wide Children's Health Assessment and Planning Survey (CCHAPS) national and state comparisons in 2008 (Power 2003).

perceived misunderstanding, or perceived sensitivity. The research team indicated their goal is to obtain a response rate of at least 33 % or 6600 completed surveys across the 6 county (Denton, Hood, Johnson, Parking, Tarrant, and Wise counties) service region in Texas.

In 2008, 7439 parents completed the survey (response rate 37 %) (Cook Children's 2009). The CCHAPS response data model (https://www.centerforchildrenshealth.org/en-us/Pages/default.aspx) created in 2010 can be analyzed from multiple angles, offering researchers a statistical data representation of decision criteria from these surveys. Since past surveys dictate the structure of the 2015 survey, the recorded statistics from the 2008 and 2012 data are used in defining the business rules for the proposed decision model.

8.2 Applying Semantic Technologies to a Survey Decision Support System

This chapter reviews the dataset for the CCHAPS and the value of creating a semantic decision support system (Cook Children's 2014). Community health survey questionnaires are targeted towards specific health issues. National- and organization-specific question banks are combined to create a comprehensive questionnaire.

There are 160 questions covering 11 major categories, also known as secondary data factors. Although the response data may not generate immediate changes in the environment, there is a great deal of interest in the data shared with academic institutions to supplement studies in specific pediatric health-care areas like abuse, asthma, dental health, mental health, obesity, safety, and prevention. Every few years, the system planning team studies and dissects the questions to assess their relevance in terms of issues of both national and local interest.

The goal of these researchers is to develop better social surveys to adapt to the changing demographics and the clinical service offerings to the target population. This chapter asks whether the inspection of questions and response criteria can provide insights into whether the correct questions are included in the survey. Preparing business rules for decision analysis is similar to preparing data for data mining. It is the point where this study suggests using semantic methods to supplement data evaluation. The following sections walk through the methods to design a semantic decision support system for the CCHAPS survey.

Analysis of Questions and Responses

The survey team wants to know whether we can rank the response values according to the question category to determine the relevancy of the question. If there is a question about asthma, there are potentially several responses with more importance in comparison to other responses. The goal is to derive semantic patterns from the question sets. For instance, we know the CCHAPS survey follows a parent–child questions model, where parent questions are required and child questions are treated as sub-questions.

There is no action on the responses to the child questions at this point in time and there is no method to draw conclusions from these sub-questions. Since sub-questions are not considered, we can classify any child question as a low priority. Risks based on cumulative or additive questions are also identified using the scale. Business rules can be created for cumulative questions, where a positive response indicates risk. For instance, if a parent indicates their child has learning disabilities, has problems with alcohol abuse, and is bullied, then these questions are considered cumulative and the child is measured as an at-risk child.

In the CCHAPS interview, the team-revealed rules were important from a data analysis perspective, but were not well documented. Many rules uncovered were substantiated from national and organizational research initiatives or slicing the data in a different way. It is important to understand why they considered looking at different combinations of data, and if documenting these relationships as business rules could reveal other patterns further down the road. The team also expressed difficulty in designing a targeted survey with the right amount of questions. People give up on the survey when questions are so specific; however, the team is unable to trim the survey for fear of losing responses. Some questions are based purely on curiosity and have no definitive drivers leading to highly probabilistic responses. Each of the 2008 and 2012 surveys reveals a gap between questions affecting actual health-care outcomes and those used for research. Documenting the business rules supporting these questions provides additional information when response data are variable as illustrated by questions 1, 2, and 5.

1. How familiar are you with the types of health-care services available in your area?
2. How familiar are you with where you can get information about health issues that affect your child?
3. How familiar are you with the types of mental health services that are available in your community?
4. How familiar are you with the types of dental services available in your community?
5. How familiar are you with the types of injury prevention services available in your community?

The questions about the availability of services do not reveal any risk, but the side-by-side comparison is useful in visualizing the responses to similar questions. At a minimum, we can suggest combining questions to reduce the repetition in the survey. All questions are, however, part of secondary data categories (demographics, physical health, oral health, emotional and mental health, safety, family activities, health insurance, and access to care), are deterministic, and provide documented business rules based on national and local statistics. Demographic secondary data, for instance, contain several indicators (birth rates, death rates, income, and household composition) can be recorded as business rules in the decision model. This information provides key variables and source details associated with specific question layouts. These variables are used to derive potential patterns independent of data mining methods. In order to understand whether the team is asking the right questions, we have to understand all the facts in the survey such as the length of the survey, the number of sensitive questions, and the responses to sensitive questions. The goal is to provide a way for the team to determine whether to remove the question, keep it, or change the wording. As part of this analysis, we wanted to determine how clearly a question is stated and flagged those questions according to ambiguity and according to the following criteria:

1. Unambiguous—yes or no response; check all that apply; distinct options
2. Slightly unambiguous—response to question with a dependency to other questions; in general type question, have a do not know response type
3. Slightly ambiguous—uses special terms, abbreviations—blank response
4. Ambiguous—response type includes the other choice or a written section
5. Strongly ambiguous—sensitive questions and answer not provided.

Although the labeling is subjective, it reveals patterns in question pairings between parent/child questions that are commonly avoided. The derived value is described by the percentage of favorable responses versus negative responses. The following categories are typically analyzed together to reveal patterns:

- Physical health and access-to-care categories are evaluated together.
- Dental, safety, and mental health categories are also evaluated together.
- Breaks in income and the Special Supplemental Nutrition Program for women, infants, and children (WIC), food stamp programs look at the federal standards for poverty limits.
- The CCHAPS survey does not address family composition questions which reveal the number of people in the household and the true family composition. These details are useful when identifying needs for reduced lunch programs in these counties.

Analysis of Secondary Data and Business Rules

The Center for Community Health Development, Texas A&M Health Science Center School of Rural Public Health, put together a secondary data report from 60 data sources covering national, state, county, and city organizations agencies (Sackett et al. 2008). They developed a set of secondary data indicators for children in the six-county service region of Cook Children's to illustrate the health of children in seven broad data categories including demographics, physical health, dental health, emotional and mental health, safety, family health and activities, and health-care access (Sackett et al. 2008).

As there are numerous factors impacting the way people respond to the survey, there are also many ways to analyze these data. A semantic model improves the documentation of business rules in an easily manageable format. A semantic model should be both measurable and testable, emphasizing both the level of the service provided by the web applications and on the end-user level (Deshpande et al. 2002). The CCHAPS team relies on the secondary data to guide the development of the survey and as their business rules. The secondary data report models the rules schema for data validation. Creating a standalone rules schema outside of a traditional database structure enables better data sharing. Survey responses do not reveal the reality of the community demographics or health history for health program decision analysis. Identifying insights into the survey questions of the most value at the community level helps to determine whether the questions will collect accurate responses. There exists a general lack of information on the developmental, mental, and physical health needs of children since the landscape of health care and health needs vary based on environmental, geographic, cultural, and social pressures. The question to ask is whether improving overall survey questions will present a better picture of this landscape, and provide more insight into asking the right questions.

Requirements and Data Analysis Techniques

How data are used really depends on the individuals using the data and the data collected have different value to different people. This section focuses on the CCHAPS committee who create surveys to capture strategic socioeconomic, demographic, and health-related data for child populations. There is a need to document business rules to document the decision criteria for the CCHAPS survey. Before we can understand the context of a survey in terms of business rules, we must understand the role of semantic quality and the options available to study a broad spectrum data model. The CCHAPS team developed a multidimensional data model to present the population statistics for the survey results. Each year is represented in the model against a secondary data category and question set. Several filters allow the user to drill down and compare decision variables between the six counties.

It is often through development and testing of web systems (including databases, services, and interfaces) that errors and gaps in requirements are discovered. For instance, the assessment team noticed that many participants skipped groups of questions related to perceived sensitivity. Their responses were blank, "don't know," or "prefer not to provide." This raises the question whether these questions were omitted purely because of their sensitive nature or whether the question wording is vague. Through traditional data mining methods, researchers can observe correlations between variables, but they cannot infer whether the survey format and questions could be analyzed and improved. The design of semantic models enables the discovery and derivation of patterns within requirements and business rules to inference a dataset for a deeper analysis.

The problem is to identify the quality of questions in terms of response choices and response data, where business rules are undocumented or do not exist. The proposed enhancement is to develop a decision process tool that allows decision makers to review business needs in order to elicit better requirements. Therefore, the need exists to determine whether a semantic solution can be justified within the scope of all other interacting systems. Although the solution is straightforward to implement, the greater value lies in the insights that a semantic model provides. The following sections describe several methods of analyzing data.

Decision Support Systems

Decision support systems applications development in business and management started to expand in the 1980s. Thereafter, natural language processing became popular in the 1990s. The literature since this time has examined manipulating quantitative models and analyzing large datasets in support of group decision making (Power 2003). Decision support systems encompass knowledge management and data mining. Both areas appear to share similar goals with semantic data management in terms of defining data and deriving patterns. In this chapter, we propose a questionnaire as decision support system. Semantic decisions derive business rules and patterns from the survey questionnaire. As data mining traditionally approaches the quantitative variables of a survey, a semantic design focuses on the qualitative variables like the actual question verbiage.

Semiautomated data mining techniques and semantic analysis produce another layer of processing to filter out variables for decision criteria. Weka and the Stanford Parser are used in parallel with the existing CCHAPS data model to analyze question syntax and derive grammatical patterns. Weka is a collection of machine learning algorithms for data mining tasks, which we leverage for question analysis. The Stanford Parser is a natural language parser that works out the grammatical structure of sentences, for instance, which groups of words go together (as "phrases") and which words are the subject or object of a verb. It is a probabilistic parser and uses knowledge of language gained from hand-parsed sentences to try to produce the most likely analysis of new sentences. Both tools do not guarantee complete accuracy, but provide metrics to help validate decision criteria, and perhaps predict business rules against grammatical patterns.

The Weka decision tree and Naïve Bayes learning algorithms were applied to sensitive questions. Weka identifies 10 of 19 questions accurately indicating a large number of sensitive questions in the dataset. In other words, the ranking rationales will need to be expressed as business rules to derive a more thorough estimate of the impact of sensitive questions in the survey model. Although the scales cover estimated sensitivity, ambiguity, and risks, the decision tree and Naïve Bayes algorithms only demonstrate an overall accuracy of 68 and 69 % which is not high. The reason for this variation is the standalone questions are very unique and do not provide an opportunity for the algorithm to learn a pattern for individual questions or for groups of questions in the same category. A detailed error analysis is required since the results are not promising from the analysis of the raw questions. Improvements in data processing can be achieved by embedding the ranking rationales into the learning algorithm.

Natural language processing techniques like dependency analysis can help to extract some semantic information, and then encode them as features for learning. The Stanford Parser uses a tagging schema to parse a sentence into typed dependencies.

How many times did this child visit an Emergency Room during the past 12 months?	If this child visited an Emergency Room for an injury during the past 12 months, how did the injury occur?	At which of the following places has this child received health care services during the past year?	Which place would you prefer to visit when your child is injured or not well?
advmod(many-2, How-1)	mark(visited-4, If-1)	prep(received-10, At-1)	det(place-2, Which-1)
amod(times-3, many-2)	det(child-3, this-2)	pobj(At-1, which-2)	dep(prefer-5, place-2)
dep(visit-7, times-3)	nsubj(visited-4, child-3)	prep(which-2, of-3)	aux(prefer-5, would-3)
aux(visit-7, did-4)	advcl(occur-21, visited-4)	det(places-6, the-4)	nsubj(prefer-5, you-4)
det(child-6, this-5)	det(Room-7, an-5)	amod(places-6, following-5)	root(ROOT-0, prefer-5)
nsubj(visit-7, child-6)	nn(Room-7, Emergency-6)	pobj(of-3, places-6)	aux(visit-7, to-6)
root(ROOT-0, visit-7)	dobj(visited-4, Room-7)	aux(received-10, has-7)	xcomp(prefer-5, visit-7)
det(Room-10, an-8)	prep(visited-4, for-8)	det(child-9, this-8)	advmod(injured-12, when-8)
nn(Room-10, Emergency-9)	det(injury-10, an-9)	nsubj(received-10, child-9)	poss(child-10, your-9)
dobj(visit-7, Room-10)	pobj(for-8, injury-10)	root(ROOT-0, received-10)	nsubjpass(injured-12, child-10)
prep(visit-7, during-11)	prep(visited-4, during-11)	nn(services-13, health-11)	auxpass(injured-12, is-11)
det(months-15, the-12)	det(months-15, the-12)	nn(services-13, care-12)	advcl(visit-7, injured-12)
amod(months-15, past-13)	amod(months-15, past-13)	dobj(received-10, services-13)	cc(injured-12, or-13)
num(months-15, 12-14)	num(months-15, 12-14)	prep(received-10, during-14)	dep(well-15, not-14)
pobj(during-11, months-15)	pobj(during-11, months-15)	det(year-17, the-15)	advmod(injured-12, well-15)
	advmod(occur-21, how-17)	amod(year-17, past-16)	
	aux(occur-21, did-18)	pobj(during-14, year-17)	
	det(injury-20, the-19)		
	nsubj(occur-21, injury-20)		
	root(ROOT-0, occur-21)		

Fig. 8.2 Parsing

Figure 8.2 is the output from the Stanford Parser. These dependencies represent the hierarchy of the text in the sentence, for instance, of the statement. To rank the value of responses, we can analyze the similarity between a question and its responses. This means if a response has higher similarity, like a question about asthma, then it would be assigned a higher importance. This approach does not need human annotations, and is easier to parse using a data mining tool or semantic definition.

Machine learning is not required for the analysis of risks based on responses to cumulative questions; however, if business rules can be identified as values for these questions, they may also be inserted into the overall processing. It is ultimately the role of the decision maker to analyze whether the combination of questions may lead to potential risks and the role of the tool to present the options. Thus, the most important thing is to define the rules to insert them in a semiautomated tool to determine derivatives.

Requirements Engineering

It is important to determine how to validate requirements as much as it is to capture them for the development. Requirements engineering is defined as the process of discovering the purpose the software is intended by identifying and documenting them in a format designed to analyze, communicate, and implement solutions (Al-Salem et al. 2007). Requirements engineering is an evolving process of eliciting process dimensions, rapid prototyping, and maintaining lightweight requirements for the life of the project. Identification and analysis are the primary goals of the requirements engineering methodology. Mandates for documentation, traceability, and change control are also an emphasis.

Requirements engineering represents the decision classification technique to derive and clean the data for analysis. This approach supports requirements quality and data exchange between processes. Requirements quality is supported by defining clearer requirements with the ability to model requirements, create relationships between business rules, and derive rules to classify requirements as semantic requirements. This begins with requirements discovery, business rules analysis, data mapping, validation, and finally process design.

In terms of analytics, there is a distinction between data and information. For instance, as a power user, you may share information via one system, but do you know if there is information cohesion between all the data points you interact with? In other words, do you have to update multiple sources to convey this information to your personal and organizational information network? Users are interested in collecting data, excited when data entry can be automated, and nervous about sharing their system with others. This results in data redundancies and gaps as different service areas collect similar data about the same patient and store it in different systems.

In the development life cycle, it is common to cradle software design, balancing needs versus wants during requirements analysis. Business rules (the constraints and conditions) are often not articulated well and may be difficult to express. If not directly associated with an application feature, rules can be overlooked. It is evident in many systems that there are a number of business rules driving the design of business models and technology. However, as in any industry, rules may be represented by the features of a tool, but seldom validated to verify whether they are consistent with current strategic goals. Technology is a means to an end, and should never be used to enforce business rules.

The first step in the requirements process is business rules discovery. This is the part of the process, where it is up to the stakeholder to state the facts and describe the systems of truth for their activities. Hold the assumption that requirements engineering should be perceived from an information perspective instead of a formal standards perspective. This means, in an ideal situation, the stakeholders all do things the same way and share the same knowledge base. However, in reality, a host of other variables including environmental, organizational, and political factors drives how the data are used and its overall quality. Therefore, involve stakeholders and review all assumptions prior to specifying business rules. In these discussions, the inflows and outflows of all the users involved in every step of the process can be documented. Tools like notation 3 (N3) or even a simplistic XML file offer a way to document these definitions, and can be used in programming later. This is the business rules context model or map that can be referenced again and again. Next, identify areas that are automated and would benefit from automation. Identify what works, what is ineffective, and what is still needed. Map it out.

Define the business rule data map. Ask stakeholders:

- What rules define the context of the process?
- Who is involved in the data flow?

- What rules are managed in every step of the process?
- What data are produced or moved through each step of the process?
- How are data moved through the process and why?

Step 2 involves prioritizing the map. With a map of rules in hand, determine the context of the questions typically answered by the user group and determine if a commercial product exists or if a customized product meets their needs. Even in the most complex environments, most end users would like to use systems that make their jobs easier. This is the step where the constraints and meaningful relationships are identified to justify the facts.

Determine the context. Ask stakeholders:

- What are the data sources?
- How do teams use the data?
- What information do individuals deliver?
- Are there redundancies?
- How are the must-haves, needs, and wants prioritized?
- Is the plan to build or buy a product?

Solutions should also be reusable in other departments with virtually the same data collection scenarios and workflows. Encourage cross-department collaboration and identify solutions reaching a broader audience. Design solutions for reuse. If strong solutions are designed or purchased, the word will spread and other groups will want to leverage the same tools where it makes sense.

Design for the majority. Ask stakeholders:

- How many applications rely on these sources?
- What teams rely on these sources?
- What security protocols are required?
- Other industry-specific questions?

Whether tools are built or purchased, the interest is for users to rely on tools adhering to standards and allow them to make decisions. The CCHAPS team recreates surveys using statistics from state and national studies, but wants to understand whether they are asking the right questions. Response data have not been substantial enough to tailor the survey and they are not comfortable acting on the findings apart from studying the data. A business rule decision support system offers a way to validate, document, and justify the pros and cons of updating survey questions. The identification of business rules is about your users and their process development workflow, where both the subject matter experts and developers are active influencers of the end design. Asking questions is a reliable requirements analysis technique to better understand requirements and define business rules for a decision model.

Visual Data Analysis

Another method to derive and understand business rules is to define a mind map, which is similar to an investigator stringing different scenarios together to understand patterns in the story. This style of brainstorming offers a very simple, visual, and tangible method of analysis. Online tools or the traditional sticky note format of writing your ideas down and putting them up on the wall allows you to move things around to determine how things group together or identify gaps. Once all the functional areas of your project area are outlined, you can begin writing down the business rules applying in each area and how they relate to each other. The only rule is writing everything down so it can generate more ideas as you stare at the pieces.

Kim (2005) demonstrates using storyboards for requirements derivation and describes that in prototyping virtual systems, requirements may be fully defined before moving to the storyboard process. This style is common in agile requirements engineering. Kim (2005) does not go into the detail of the steps to categorize or organize requirements, but lists them as points to ponder in requirements modeling. Gerard touches on a concept called abstract formalization lending well to semantic documentation or visually modeling a system (Kim 2005). Similarly, use Unified Modeling Language (UML) or Systems Modeling Language (SYSML) to model large and complex systems (Ober et al. 2011), and provide semantic optimization. Supplementing requirements and user scenarios with semantics does not translate requirements into semantic requirements; however, defining an ontology model can provide a semantic context for large systems.

Wire framing is another visual analysis tool to identify business rules and requirements for a development effort. The benefits lie in how well you define the details through annotations and descriptions. For instance, a simple web form wireframe can be detailed in terms of each item on the page using numbered markers. The logic for the form can also be translated into a semantic fact and rule table for decision modeling, creating a pseudo-prototype with clickable shapes to demonstrate the expected functionality. Consider the many options to manage requirements and business rules and the methods used to trace these items throughout a project. The goal is to create a semantic syntax for the domain model. This design works best if every element on the wireframe is labeled and described in detail. This method of analysis presents a visual platform to review with the customer, raises process questions, and allows analysts to document the rules directly to the context they are applied.

SysML, syntax of UML, was designed by systems engineers to model systems of systems with an emphasis specification, analysis, design, verification, and validation. The OMG SysML group created a Visio template that contains several shapes that define the requirements and semantic syntax for a system description. In addition to requirements diagrams, traditional UML and alternative parametric and package diagrams are available for detailed systems design. SysML provides system-level notation to qualify relationships in the architecture. Annotating requirements in a model can be handled by the semantic query notation defined by (Corby et al. 2006), or applying a general modeling language like Systems Modeling Language (SysML) (Holt 2006; Wrycza 2011). SysML is an extension of UML that is tailored to represent the various views of a system and provide a behavioral model for requirements design. It is a standard for designing a system model, but should not be considered a new modeling language that is only meant for modeling software systems. SysML leverages the UML definitions. From a requirements standpoint, SysML provides the standard language elements lending itself to validation, verification, and testing which a systems engineer would have to ascertain and potentially create using the UML modeling methods (Holt 2006; Wrycza 2011).

Business Rules Analysis

The importance of data collection lies in analytics. How data will be analyzed drives the data that are collected. A pure transactional system offers storage, access, and retrieval of data, but requires extensive manipulation to provide cross-functional metrics across systems. Most of the database systems encountered are used for data collection, and rely on web services to provide data visualizations or results-based reporting. The data warehouse technology is there to extend analysis if it adds value to your organization's data strategy. Use both formal and informal business rules to identify the data you are collecting. Formal rules exist as data drivers. Formal rules appear in applications, systems, and processes that have been put in place to collect this specific information. Formal rules also appear as master data, the people, places, things, and concepts that drive an organization's decisions. Every patient interaction with an organization such as face-to-face visit, social media, phone call, and website interaction is master data. The CCHAPS survey relies on these interactions to identify the target audience for its survey. If the data collected are not used to communicate further information regarding the patient or health-care delivery, its value decreases and becomes stale over a period of time.

Informal rules exist in the data itself, and this is the data people manage via personal productivity tools like knowledge managers, spreadsheets, surveys, and other forms external to master data collection. Informal rules may represent hidden rules or data that have grown so much it warrants transition into a structured process. Review data patterns and derive rules about the data where rules do not exist or no longer fit the context of the data domain. Determine who, what, and why these data are collected and how they are collected. It often turns out that these informal processes were created to fill a gap where fields did not exist to capture the data in the primary system.

As time passes, both the data and the value of the data grow. As departments begin using these formal and informal systems more and more, they create a new business justification for support and come across scenarios where other teams can also benefit from using this custom solution. Healthcare systems are unable to keep up with the advancements in technology which becomes difficult to manage as data grows. This is a side effect from the rate which technology changes and health-care systems can adopt new technologies. Often replacing the custom solution with an updated or alternative system is viable and immediately addresses the data collection process across multiple groups. Still, in other cases, a more intense redesign may be required to merge the data with a primary system or bring it up-to-date.

Semantic Analysis

Resource description framework (RDF) is a standard model to organize requirements semantically into term-sets called triples for data interchange on the Web (Muoz et al. 2005). It is basically a data format based on graphs to decribe resources on the Web. This traditional web data model can be applied to a non-web context using web Uniform Resource Identifiers (URIs) to name relationships between things known as triples. Whether web URIs are available or not, the format of the data model allows business users to classify the context of their data domain in a succinct format outside of a relational database. Triples take the form of semantic statements ordered by subject, predicate, and object. Essentially, the definition of triples allows anything to become definable, by specifying a semantic framework for that thing. A file of triples for a specific thing may standalone as a query-able database or may be linked with other triple files to build complex relational descriptions about many things.

The process of defining a requirements model semantically generates meaningful patterns, rule-sets, or logic to build value into the data. In the triple (ab:craig ab:email ?craigEmail), the subject stands as the resource identifier for the statement and must be qualified with a property description or predicate. The predicate may include a property value, but is not mandatory to define the statement, although it adds semantic quality to the syntax described. The subject and predicate must belong to the same namespace to prevent confusion between similar terms. This statement returns any e-mails associated with the name craig; however, if there are (Muoz et al. 2005) multiple unique individuals with the name craig (e.g., craig 1 and craig 2) all e-mails associated with all craig's are returned. A file created with rules about this individual offers a view about everything related to this individual that can be queried. Semantic query languages like SPARQL (DuCharme 2011), a query language for RDF (Muoz et al. 2005), and SPIN (Fürber et al. 2010), a syntax for SPARQL, are the semantic inferencing tools utilized in specifying and solving problems arising in big datasets and simpler scenarios. SPARQL is a graph-matching query language (Lawrynowicz 2014). These rules are implemented using SPARQL CONSTRUCT, or SPARQL UPDATE requests, and constraints are specified using SPARQL ASK and SPARQL CONSTRUCT queries or corresponding SPIN templates (Knublauch 2014).

8.3 Pattern Analysis and Validation Techniques

Pattern discovery is a data mining construct, and SPARQL queries are the defined patterns governing the dataset. As the contribution of semantic work and datasets increases, the research into the possible learning algorithms, data patterns, and predicitive modeling is increasing. Given a dataset, the space of patterns is searched systematically from most general patterns to more specific ones (Lawrynowicz 2013). The triple pattern is the foremost dependency pattern in SPARQL. Within the triple pattern is best to avoid the query analysis against definitions of vocabulary terms which offer little instance data (Hartig 2012; Smirnov et al. 2004). This rule is important to consider as we define statements from the Stanford Parser analysis since there is the opportunity for repetition from these statements that would add processing time to the evaluation.

The RDF dataset organizes the business rules for the CCHAPS survey into specific data collections according to the secondary data categories and into named graphs according to the survey versions. The RDF triplicate form is a series of statements producing a graphical representation of the relationships between resources known as a triple store. When one or more triples are grouped and organized together via an additional identifier, this is known as a named graph or quad store (Cambridge Semantics 2014). Associating triples into a named graph enables working with a dataset as a whole rather than as individual statements and expanded querying in SPARQL. In the CCHAPS example, we can describe secondary data categories in terms of their busines rules, or update and delete the whole group (Cambridge Semantics 2014). There are several named graph patterns to handle data management in RDF according to the structure of the dataset. For the CCHAPS survey versions, each version can be interpretted as a graph or the secondary data categories according to their question sensitivity, and can be graphed individually. The patterns will vary based on the scope of the data in each graph (Cambridge Semantics 2014). The data model is designed to fulfill the following core requirements according to named graphs definitions provided by research (Carroll et al. 2005):

- Representation of meta-information—The model should allow a more efficient representation of meta-information than the RDF reification mechanism.
- Unique identification of RDF data—The model should provide a mechanism for globally unique identification of RDF data, so that different information providers can express meta-information about the same RDF data.
- Backward compatibility—In order to provide as much backward compatibility with existing RDF data and deployed applications as possible, the design should keep close to the RDF recommendations.
- Exchange of meta-information—The model should be accomapanied with syntaxes for publishing and exchanging information together with meta-information.

Tracking the Source of Triples in a Dataset Graphing the source of a dataset is the most query-able pattern in SPARQL. It allows the verification of validation of the original source and the appended sources or rules that comprise the dataset. The key is creating relationships against well-understood rules that define the content of a dataset. This is the best pattern to use when provenance data are important (Carroll et al. 2005).

Organizing the Business Rules for a Specific Data Category To manage the secondary data for individual surveys, consider the option to graph per resource. The benefit of graphing per resource is that information about both the secondary data and the source qualifier/business rules can be stored in the same data store.

Adding Notes or Derived Data to a Collection of Triples Annotating a graph is useful to capture the description of a graph. It can include whatever information may be useful in defining a graph. Like organizing the data per resource, annotation treats the data collections as a whole instead of as individual statements.

8.4 Architecture of a Semantic Decision Domain

Semantic ontologies are working models of requirements. Defining the business rules for a domain in a working ontology allows requirements to be validated and tested. The domain context also provides traceability since the framework defines the relationships between the rules and the implementation that would otherwise be difficult to infer. It is promising to investigate the representation of requirement knowledge and information in process reengineering. Exploring inference rules for requirements modeling, analyzing, and reasoning are also interesting. (Yanwu et al. 2008) suggest exploring semantic mapping between different ontology representations, and agree that consistent representation of requirement knowledge and information could combine advantages from different techniques, support different kinds of development processes, and, to a large degree, contribute to reuse and traceability.

The example discussed in this chapter focuses on the design of a decision support system. Models provide decision makers a litmus test to verify and weigh the pros and cons of a choice. For the CCHAPS survey model, this focus is on the questions in the survey, and whether they advance the understanding of the pediatric health-care environment in north Texas. Secondary data categories represent the majority of the validation criteria represented in the survey, but the proposed data model also considers question sensitivity uncovered by newer research. The subsequent sections demonstrate the new decision process to evaluate questions according to natural language formatting, the organization of questions in the survey, and the perceived sensitivity of the subject matter.

Conceptualizing the domain ontology for the CCHAPS survey creates a snapshot of all the instances a question or secondary data are used across the survey. A key issue with the survey today is its length and question redundancy. The domain ontology as a decision tool enhances the analysis methods and provides a working knowledge base to specify the rules that guide the maintenance of future surveys. Researchers can query facts and references regarding the survey to supplement their decisions. The tool itself does not generate a survey or provide new questions, but reveals patterns in the natural language and decision criteria that are not apparent by processing the response data alone. A clearly described requirements model is easy to interpret and traceable to a business rule. A strongly defined model is an essential component for the requirements definition and lends itself to building clear business cases from business rules. (Corby et al. 2006). support the idea that ontologies are the foundations of the semantic web and the keystone of the web-automated tasks: searching, merging, sharing, maintaining, customizing, and monitoring. These

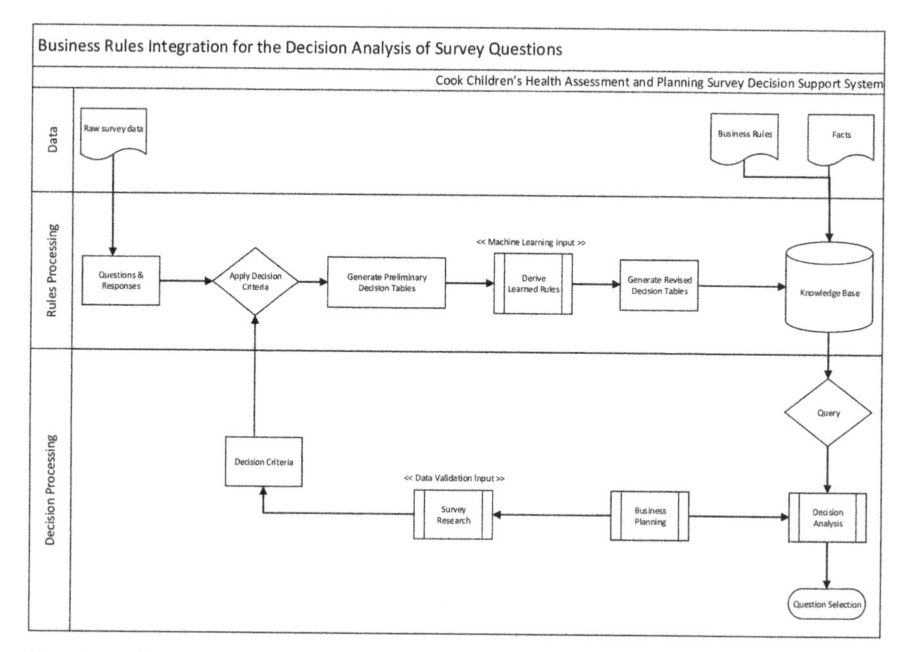

Fig. 8.3 Business rules

ideas are the cornerstone of a semantic decision support system as well. (Corby et al. 2006). present a simple semantic notation for a query that can be built into the semantic model defined within the scope of business rules. The semantic query is essentially built from a resource requirement. A web resource requirement is required to formulate a query. Similarly, a series of statements describing the survey are considered requirements for the survey resource.

The decision model, Fig. 8.3, is presented as a business rules knowledge base consisting of a data layer, rules processing layer, and decision processing layer. It requires the subject matter experts and decision makers to understand the survey and the secondary data context for the CCHAPS domain. It requires the input of new decision criteria/requirements, business rules and facts into a sentential structure. Semiautomated tools are used to preprocess and classify the requirements. Then, we utilize the triple format (subject, predicate, and object) to define the survey domain ontology and specify the business rule associations. The survey question equals the subject, the ranking rationale equals the predicate, and the business rules represent the object for each statement. SPIN, also known as SPARQL rules, can be used to calculate the value of properties based on other properties and isolate a set of rules to be executed under certain conditions (Fürber et al. 2010).

These variables are stated in terms of compound semantic statements for the survey domain. Both simple and compound semantic statements provide implicit declarations. For instance, the rules for the question, "How often do you talk to this child about drugs and alcohol?" can be defined as both a simple and compound triple statements depending on the number of response choices and the number of secondary topics the question is related to. Therefore, a question about drugs and alcohol can be related to questions about healthy behaviors, mental health, family activities, and parent profiles. We can look at the number of questions presented in each area for these topics and review overlaps in content and wording in addition to the varibility in the responses received. Each predicate and object variation illustrates the compound relationship to the subject considered relatable by the business rules context. In the named graph structure, these compound statements enable recursive query expressions relating similar statements together. It is important to note that the use of such recursive functionality can add cost to performance time. The use of *from* requires a union between graphs, whereas the use of *graph* in queries limits lookups to one graph at a time.

8.5 Conclusion

As health-care organizations have made the leap into the digital age, they are data centric, capturing all kinds of data using many customized tools. Questions exist about data relevancy and whether these data gathering methods are encroaching on privacy, but are health-care providers really data aware? In other words, the focus needs to shift from data collection towards data quality and understanding the value of data. Semantic quality is the value attached to the meaning of a requirement. For instance, a business rule conducive to the operation or function of a system can be assigned a priority or dependency. Stating priority or dependency identifies the value of this rule in terms of the overall operation of the system.

Decision criteria are the requirements for the semantic decision support system, and the requirements identification process can be likened to the evolution of the Web which has inched from a documentation focus to an application focus, and is now migrating to an information focus. Survey researchers and statisticians are interested in deriving relationships from the data collected. In terms of analytics, there is a distinction between data and information. This distinction can be expressed in terms of the explicit decision criteria, the requirements for the survey, but they can also be implicitly expressed in terms of business rules. Business rules (the constraints and conditions) are often not articulated well if not directly associated with a program feature and can be overlooked. Rules may be represented by the decision criteria, but are seldom validated to verify whether they are consistent with current strategic goals. Therefore, the first step in the requirements process is business rules discovery. Requirements engineering when perceived from an information perspective instead of a formal standards perspective offer users a broader view of data relationships. There are three immediate benefits to developing semantic models including capturing the outline of interactions between platforms in a system to prevent solution development in data silos, the identification of requirements gaps to reduce semantic debt, and the derivation of business rules datasets.

Creating semantic models in requirements engineering explores whether the design of a semantic decision support system can provide validation support, identify requirements, and reduce requirements overhead contributing to semantic debt. The model acts as a decision support system to allow analysts to view interactions between systems allowing them to justify the need for new development and pinpoint issues appearing in legacy systems. Since dependence on these legacy tools is defined to be critical according to their role in the workflow, replacing or changing these systems must be carefully considered. The discovery of business rules or data quality issues using the decision support system can help analysts plan projects if the design works as intended. The analysis of semantic techniques is still in a growth phase, and novel approaches in applying these techniques for web development and data transformation are valuable areas to research since they are not yet generally understood by mainstream data analysts. Using semantic models in requirements engineering and development offers an opportunity to test the value of these techniques in development today.

References

Al-Salem, L. S., & Abu-Samaha, A. (2007). Eliciting web application requirements—an industrial case study. *Journal of Systems and Software, 80*(3), 294–313.

Cambridge Semantics. (2014). Data management design patterns. http://www.cambridgesemantics.com/semantic-university/data-management-design-patterns. Retrieved June 2014

Carroll J., Bizer C., Hayes, P., Stickler P. (2005). Named graphs. In Web Semantics: *Science, Services and Agents on the World Wide Web*. (Vol 3, pp247-267)

Cook Children's. (2009). Cook Children's Health Care System Community-wide Children's Health Assessment and Survey (CCHAPS)Technical Appendix (p.7) https://www.centerforchildrenshealth.org/SiteCollectionDocuments/TechnicalAppendix.pdf Retrieved April 2014.

Cook Children's. (2014). Center for children's health community-wide children's health assessment & planning survey (CCHAPS). http://www.cookchildrens.org/AboutUs/CHO/CCHAPS/Pages/default.aspx

Corby, O., Dieng-Kuntz, R., Gandon, F., & Faron-Zucker, C. (2006). Searching the semantic web: Approximate query processing based on ontologies. *Intelligent Systems, IEEE, 21*(1), 20–27. doi:10.1109/MIS.2006.16.

Ducharme, B. (2011). *Learning SPARQL*: Querying and Updating with Sparql 1.1. Sebastopol, CA: O'Reilly.

Fowler, F. J. (1988). *Survey research methods* (Applied Social Research, Vol. 1). Newbury Park: Sage Publications.

Fürber, C., & Hepp, M. (2010). Using SPARQL and SPIN for data quality management on the semantic web. In W. Abramowicz & R. Tolksdorf (Eds.), *Business information systems* (Vol. 47, pp. 35–46). Springer: Berlin Heidelberg.

Hartig, O. (2012). *An introduction to SPARQL and queries over linked data*. ICWE.

Holt, J. (2006). SysML: Describing the system. *Information Professional, 3*(4), 35.

Kim, G. (2005). *Requirements engineering and storyboarding designing virtual reality systems the structured approach* (pp. 14–26). Springer: London.

Knublauch, H. (2014). SPIN - Modeling Vocabulary. W3C Member Submission. http://spinrdf.org/spin.html

Kramer, J. K. A. M. (2011). *Nonprofit management collective impact* (Vol. 63). Standford Social Innovation Review. http://www.ssireview.org/articles/entry/collective_impact

Kreuter, F., Presser, S., & Tourangeau, R. (2008). Social desirability bias in CATI, IVR, and web surveys. *Public Opinion Quarterly, 72*(5), 847–865. doi:citeulike-article-id:10551970; 10.1093/poq/nfn063

Lawrynowicz, A., & Potoniec, J. (2014). Pattern Based Feature Construction in Semantic Data Mining. *International Journal on Semantic Web and Information Systems* (IJSWIS), *10*(1), 27–65. doi:10.4018/ijswis.2014010102

Mack, N., C. Woodsong, K. MacQueen, G. Guest, and E. Namey. (2005). Qualitative Research Methods: A Data Collector's Field Guide. Research Triangle Park, NC: Family Health International.

Muoz, S., & Gutierrez, C. (2005, 31 Oct–2 Nov 2005). *Interpretations between RDF and the logical data model*. Paper presented at the Web Congress, 2005. LA-WEB 2005. Third Latin American.

Ober, I., Ober, I., Dragomir, I., & Aboussoror, E. A. (2011). UML/SysML semantic tunings. *Innovations in Systems and Software Engineering, 7*(4), 257–264. doi:10.1007/s11334-011-0163-2.

Power, D. J. (2003). *A brief history of decision support systems*. doi:citeulike-article-id:2032414.

Sackett, S., Burdine, J. N. & Wendel, M. L. (2008). Secondary data report: Texas A & M Health Science Center School of Rural Public Health.

Smirnov, A., & Chiueh, T. (2004). *A portable implementation framework for intrusion-resilient database management systems*. Paper presented at the Dependable Systems and Networks, 2004 International Conference on 28 June–1 July 2004.

Wrycza, S., Marcinkowski, B. (2011). SysML requirement diagrams: Banking transactional platform case study. *Research in systems analysis and design: Models and methods* (Vol. 93, pp. 15–22): Springer: Berlin Heidelberg.

Yanwu, Y., Fen, X., Wensheng, Z., Xian, X., Yiqun, L., & Xuhui, L. (2008). *Towards semantic requirement engineering*. Paper presented at the Semantic Computing and Systems, 2008. WSCS '08. IEEE International Workshop on 14–15 July 2008.

Deshpande, Y. S. M., & Ginige, A. (2002). Web engineering. *Journal of Web Engineering, 1*(1), 4–14.

Chapter 9
University Ontology: A Case Study at Ahlia University

Karim Hadjar

9.1 Introduction

Nowadays, the web has shifted into another dimension: the semantics. The language of the web, HTML, has embraced semantics with version 5. Almost all of the fields have their own ontology. Since I have been an academic for over two decades, I have learned that there are few serious developments in terms of university ontologies, which led me to build one.

Several tools and methods have been developed to build ontologies. Rather than focusing all the attention on information, we also focus on the core concepts in using the ontology and its relationships. The most well-known and widespread tool for editing and developing ontologies is Protégé. Its graphical user interface (GUI) lets the developers concentrate on the concept rather than thinking about the syntax of the output language. Protégé has a pliable data design and extendible plug-ins. In this chapter, the definition of the university concept is clarified through a university ontology. Creating a university ontology with Protégé is the objective of this chapter.

Ahlia University is taken as a case study for the development of the ontology, and several phases are outlined, e.g., superclass, hierarchy of subclasses, creating subclasses instances, retrieving queries, graphs, and visualization views. The case study is limited to few departments and courses, as an example. This implies that since the model works for one university, it will work for other universities, with minor changes.

This chapter is organized as follows: In the next section, I discuss the steps required for building the university ontology. The following section presents my case study: "Ahlia University." Next, some queries are applied and exposed on the Ahlia University ontology. Finally, I conclude this chapter with a discussion.

K. Hadjar (✉)
Multimedia Department Chairman, Ahlia University, Manama, Bahrain
e-mail: khajjar@ahlia.edu.bh

M. Workman (ed.), *Semantic Web*, DOI 10.1007/978-3-319-16658-2_9

9.2 University Ontology

There are many data shared between organizations and organization divisions, which can be used in building up the university structure. Yet there did not seem to be a suitable shared terminology for presenting such information in linked data. Based on the need of a clear structure for any organization, I have developed university ontology as the basic structure model to share between all organizations who do not like to start from scratch. In fact, a lightweight, highly reusable ontology, which did not try to model particular organizational structures, is required.

Building the Ontology

To increase the ontology efficiency, I need to ensure that ontology is defined as a formal specification, explicit and consensual conceptualization of a domain (Guarino 1998). Definitely, the development and design of ontology helps people to recognize and answer the questions about domains (Ghorbel et al. 2008). It comprises a group of concepts related together in an organizational method. In this chapter, I focus on specialized ontologies, that is, domain ontology and task ontology. These are reusable ontologies within a given domain, but not from one domain to another, while all tasks performed in a given domain are within the ontology (Guarino 1998). According to Mizoguchi et al. (1995), the ontology task is to describe a curtain vocabulary related to a task. The reuse of ontology is critical. I have to build the ontology from scratch by following a known methodology.

Ontology Development Methodology

To ensure the consistency of ontology structure and to increase its efficiency during development, I have followed the guidelines from many sources. First, I have studied how to build ontology by using a guideline from Noy and McGuinness (2001) and Zhao et al. (2004). The guide was built using Protégé ontology editor, which is the same tool that I have used for the university ontology development. I have studied a few ontology development methodologies, and finally I have decided to follow a recently defined methodology from Gruninger and Fox (1995), incorporating with the guide from Mizoguchi et al. (1995). This ontology development covers the steps from the initiation phase to the data retrieval phase of ontology. Specification and conceptualization (Gruber 1993) are the two main steps available in this methodology. Obtaining knowledge about the domain is the objective of the first step. Moreover, organizing, i.e., structuring the information using exterior demonstrations independently from the environment and implementation language is the second objective.

Specification

The scope puts boundaries around the ontology; requiring defining what has to be involved. This step was suggested for an advance stage in the ontology development: A guide to create the first ontology (Guarino 1998) is included at this stage to minimize the process of analyzing concepts and data, particularly for the range and difficulty of the university model ontology. During the iterations for following the verification, the process will be adjusted when needed. I have considered the needs for elaborating the university structure project with theories related to higher education organizations. It is the first prototype, and the considered concepts are not related to all divisions in an organization. Therefore, it includes general concepts for the university abstract model.

Previous domain analysis was necessary to be done as the first step. In this work, the presentation for framing the university structure and the relevant documents were collected from a number of organization charts of different universities. Furthermore, advice from management leaders of universities and faculty were taken into consideration. The Gruninger and Fox methodology point of view was taken into account. Problems arise when people need information but the systems do not provide it. The motivation scenarios are followed. In addition, templates have been used in order to define motivation scenarios and link them to the people involved. A set of solutions to all problems is made available whenever the semantic features can be resolved.

Conceptualization

In this step, the terms used in representing the most important entities in the university structure are enumerated as classes shown in Table 9.1. Definitions of the main classes are listed after the table. All the concepts appearing in the figure mostly focus on the main departments in any University, e.g., AcademicAffairs, AdministrativeAffairs, President, Deans, Chairs, Faculty, Student, Courses, Library, Gym, WebSite, BookStore, etc.

Table 9.1 Key item list as class and subclass

Class	Subclass
Courses	Graduate courses
	Under graduate courses
Programs	Bachelor program
	Master program
	PhD program
Person	Employee
	Students

Table 9.2 Relation between the university model classes

Class name	Relation	Class name	Inverse relation
Department	Has head	Chair	Is head of
Students	Take	Courses	Are taken by
Faculty	Publishes	Articles	Is published by
Classes	Attended by	Students	Attends

- The class “Courses” is defined as the basic two categories of the courses available at most universities, “GraduateCourses” and “UnderGraduateCourses.”
- The class “Programs,” defined as the possible offered programs by universities at their start or at later stages of their program. And the three subclasses are, “BachelorProgram,” “MasterProgram,” and “PhDProgram.”
- The class “Person” divides the people at the university to two types: Employees (staff and faculty) and Students.
- The main relations, attributes, and properties have been created as shown in Table 9.2. Figure 9.1 contains the object properties according to the relationship, which I want to add between the classes “Professor” “advises”, and the courses “areOfferedBy.”

Table 9.3 illustrates the relationship between individual to data literal, for example, the Course has “CRN,” “coursename,” “creditHours.”

In a general usage, a restriction can be a general form of instructions that sets a limited border defined for a function or a type of process. These relations were captured in a semantic diagram to represent the relations between components.

Property and Relationship

Since having only classes cannot answer all the enquiries, defining links inside or between the classes is needed (such as properties). In the example, I have used property, which shows a relationship between individual and individual, a relationship between Individuals at university ontology, such as property, and a faculty as an advisor of student. I have also defined Object Properties, Domain, and Ranges, for example:

```
<owl: Object Property rdf: about = advisor>
<rdfs: domain rdf: resource=student/>
<rdfs: range rdf: resource=faculty/>
</owl: object Property>
```

In the top layer of the university ontology there is: “Person,” “course,” “committee,” “AcademicAfairs,” “Admission,” “University,” etc. In the middle layer of the university ontology there is: “AdminStaff,” “Student,” “articles,” “books” and

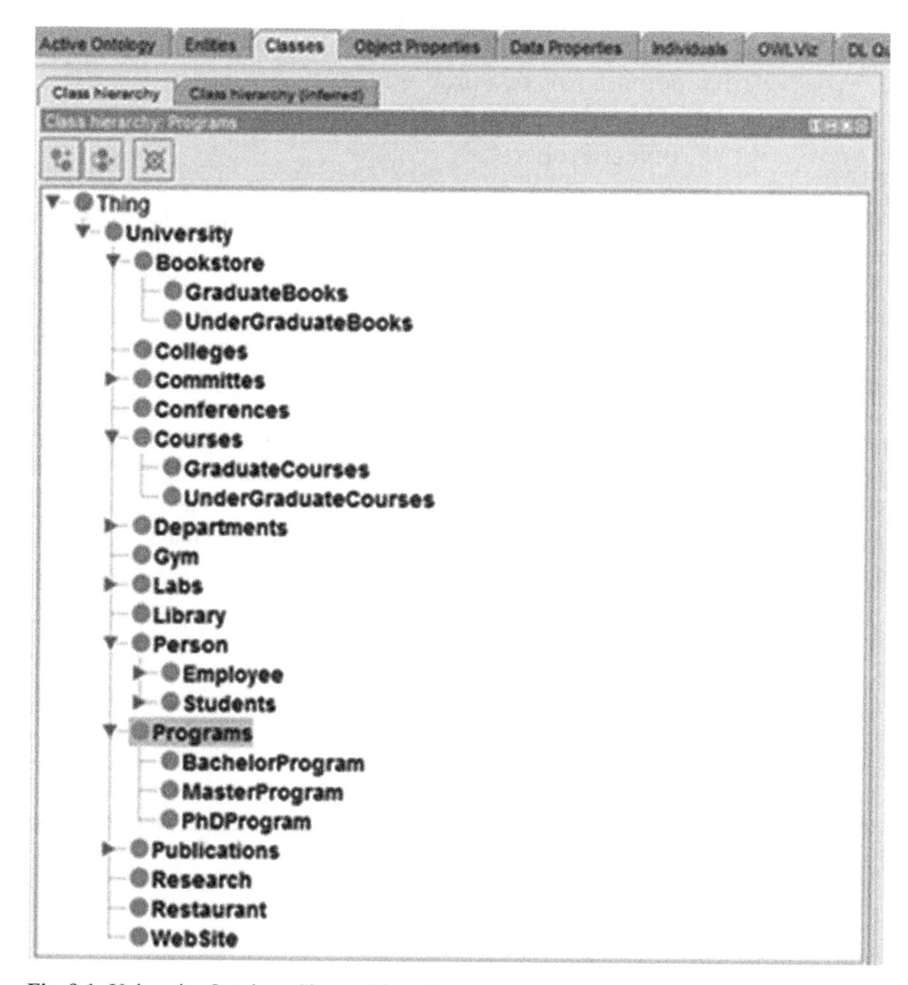

Fig. 9.1 University Ontology Classes Hierarchy

Table 9.3 University ontology object property

Class	Property
Faculty	Email address
	Mobile
Course	CRN
	Course name
	Credit hours

"subject," "library," "colleges" and "departments," etc. And the bottom layer includes: "Chair (Professor)," "Teaching Assistant," "Dean," "Director," "Visiting Professor" and "Professor Types," etc., for example:

```
<owl: object property rdf: about = TeacherOf/>
<rdfs: domain rdf: resource = Faculty/>
<rdfs: range rdf: resource = Course />
</owl: object Property>.
```

The object Property—TeacherOf; its domain is in Faculty and range in Course. It means that TeacherOf Property value will be only just opposite to the isFaculty property because has Property is always inverse to is Property. The relation of Inclusion (rdfs: subPropetryof), equivalent (owl: equivalentPropetry), and Inverse (owl: inverseOf), and the limitation of function (owl: FunctionalPropetry) and inverse function (owl: InverseFunctionalProperty).

Since the conceptual model of the ontology has been created, the next step is to define related instances. For each instance, I have described: a name, the name of concept it belongs to, and its attribute values. The instance (individual) is described first, then the right class was selected, and finally its instances for the class are created. Use rdf: type to state its class, and one instance can belong to many classes or many class belongs to same instances, for example:

```
</owl: thing rdf: about= AdvancedDatabaseSystems >
<rdf: type rdf: resource= #course/>
<rdf: type rdf: resource= #student/>
</owl: thing>
```

Here it defines an individual or instance AdvancedDatabaseSystems, which belongs to the class "course" and "student." In which rdf: Type has appeared twice, it shows that this instance belongs to two classes at the same time.

Implementation

I have chosen Protégé 4.1 (Protégé 2013) in order to implement the ontology, due to its extensibility, quick prototyping, and application development. Protégé ontologies are easily exported into different formats including Resource Description Framework (RDF) schema, Web Ontology Language (OWL), and Top Braid Composer (2013), which I have used at later stages in querying. Particularly, I have implemented the university ontology in OWL. Structured relations are transformed into bidirectional relations while modeling in OWL. Moreover, only relations that are necessary in answering competence questions were modeled in ontology.

Verification

Consistency validation and classification are verified by using the Reasoner. During the process of charging classes and attributes, I have used incremental and continuous verification to avoid future propagation errors. In the Reasoner, any class which is unsatisfiable is shown in red color indicating that error exists. At this point, it is very important to see how classes are defined (disjoint, isSubclassOf, Partial Class, Defined Class, etc.) and how are their restrictions (unionOf, allValuesFrom, etc.). Classification process is either for the whole ontology or for selected subtrees only. When the test is completed, the whole ontology, errors were listed, moving from bottom- to upper-level class. To compare the ontology execution with its conceptualization, graphs were generated using OWLViz (Graphviz 2013) and OntoViz plug-ins (OWL Web Ontology Language Overview 2013).

9.3 Case: Ahlia University Ontology

Ahlia University ontology defines elements to describe Ahlia University and its activities, which can occur between Departments, Faculty, and Students. I have built the ontology based on the organization chart available on the university website, and all data used for testing my work were also taken from the catalogue available. Since my base of the ontology was the university ontology, I only had to make some changes on the classes following the organization chart. Concepts (classes) such as: Departments, Degrees, Deans, Chairs, Faculty, Student, Courses, Library, CareerCenter, WebSite, ICTCenter, Labs…etc., More relations (rules) were added between the classes to show how they are related and linked to each other. The Ahlia University Ontology also includes relationships between classes. For instance, the relationship "teaches/isTaughtBy" is between Faculty class and Courses class. Other relationships are added, such as: hasHead/ isHeadOf, hadMember/isMemberOf, etc.

As shown in figure 9.2, some of Ahlia University-related classes and subclasses are listed. All the concepts appearing in the figure are mostly focused on the students, faculty, and course based:

- The class "AhliaUniversity" is the highest-level class in this domain.
- The class "Assistant" defined as the basic two categories of the position of an Assistant available at most universities, ResearchAssistant and TeachingAssistant.
- The class "Professor," defined as the rank type of the faculty at the universities. I have listed its subclasses (AssistantProfessor, AssociateProfessor, FullProf and VisitingProf) (Fig. 9.2).

Query Retrieval

A very powerful feature tab is available in Protégé, which is the description logics (DL) query. Considered as one of the basic plugins in Protégé 4, and is either available as a tab or a widget. It is based on the Manchester OWL syntax, which is a query language supported by the plugin, and a user-friendly syntax for OWL DL.

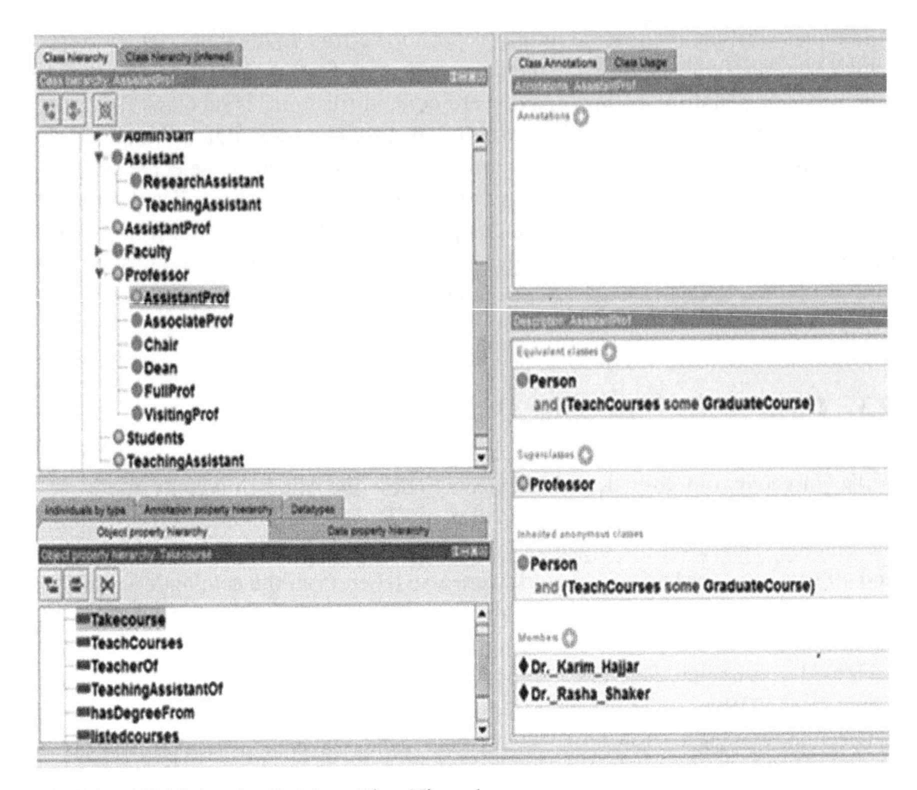

Fig. 9.2 Ahlia University Ontology Class Hierarchy

A frame which is fundamentally based on the information is collected about a specific class, individual, or a property, into a single construct (Noy and McGuinness 2001). Here again, the query retrieval process has gone along the steps depicted in previous sections, and illustrated in Figs. 9.3 and 9.4.

DL Query 1

- Which courses does Dr Karim teach?
- Courses_Offered and CourseFaculty value Dr_Al-Hadjar_Karim

DL Query 2

- The list of available faculty on Saturday
- Course_Faculty and available value Saturday

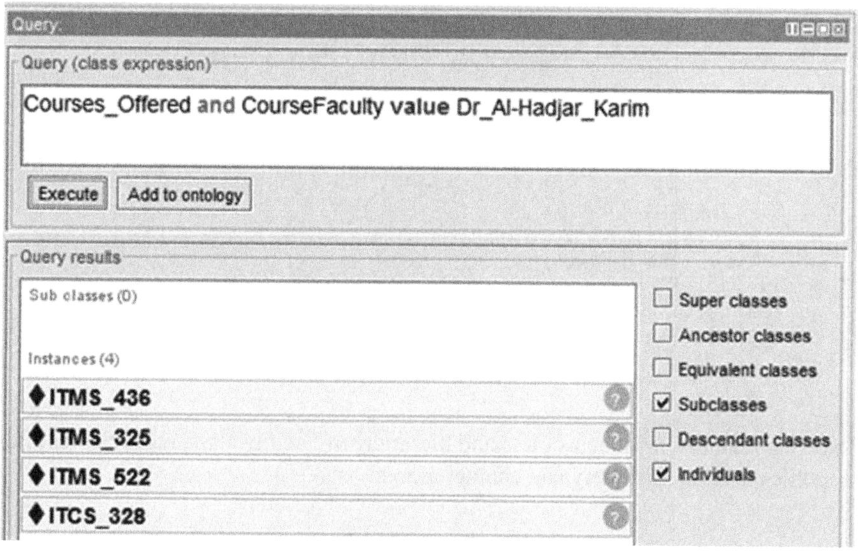

Fig. 9.3 Snapshot of the description logics (DL) query 1

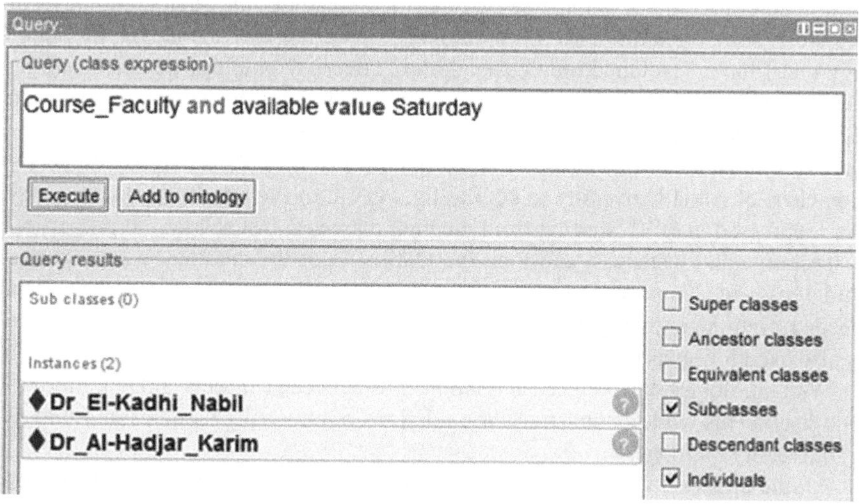

Fig. 9.4 Snapshot of the description logics (DL) query 2

Below, I have listed example of the data retrieved from the ontology using Special Protocol and RDF Query Language (SPARQL) query editor, available in TopBraid Composer.

Query

```
PREFIX rdf:<http://www.w3.org/1999/02/22-rdf-syntax-ns#>
PREFIX uo:
<http://www.semanticweb.org/ontologies/2011/9/Ontology13195718561
22.owl#>
SELECT? X, ?Y, ?Z
WHERE
{?X rdf: type uo:Student
?Y rdf: type uo:Faculty
?Z rdf: type uo:Course
?X uo:advisor ?Y
?Y uo:teacherOf ?Z
?X uo:takesCourse ?Z }
```

Aside the features of Students class and hierarchy of Faculty class, most classes and properties used in this query can characterize it.

Conclusion

This chapter presents my contribution relative to creating a university ontology. All the work done on Ahlia University ontology was a reuse of the university ontology that I have developed previously. Ahlia University ontology describes all the departments under the university structure and the relationships that exist between them. For this chapter, I have shown my modified OWL version of the university ontology and added more classes and restrictions based on the university organization chart of Ahlia University to get the final OWL of the ontology. The ontology was expressed in OWL starting from creating classes and subclasses to properties, restrictions, and instances, and then the OWL file of the ontology was imported into TopBraid Composer for more powerful data retrieval software, to get the data needed from the ontology easily with short SPARQL queries. DL query in Protégé is also used for querying.

With ontologies, the focus is on relationships between concepts and not information itself. This work demonstrates the relations of university modules in the form of university ontology.

All the attention was given to the core concepts of using ontology and its relationships rather than information. As a future work in the domain of ontology in higher education, one can consider the following topics. The list may include but is not limited to:

- E-learning applications ontology
- Ontology sharing and reuse
- Graphs ontology
- Enterprise ontology
- Ontology matching and alignment

References

Ghorbel, H., Bahri, A., & Bouaziz, R. (2008). Les langages de description des ontologies: RDF & OWL. Acte des huitièmes journées scientifique des jeunes chercheurs en Génie Electrique et Informatique (GEI), Sousse-Tunisia.

Graphviz. (2013). Graph visualization software. http://www.graphviz.org. Accessed 4 Jan 2014.

Gruber, T. R. (1993). A translation approach to portable ontology specifications. *Knowledge Acquisition, 5*(2), 199–220.

Gruninger, M., & Fox, M. S. (1995). Methodology for the design and Evaluation of Ontologies. IJCAI Workshop on Basic Ontological in Knowledge sharing, Montreal, Canada.

Guarino, N. (1998). Formal ontologies and information systems. In proceedings of FOIS'98, IOS Press, Amsterdam.

Mizoguchi, R., Vanwelkenhuysen, J., & Ikeda, M. (1995). Task ontology for reuse of problem solving knowledge. In proceedings of the 2nd International Conference on Building and Sharing of Very Large-Scale Knowledge Bases. (KB & KS'95).

Noy, N. F., & McGuinness, D. L. (2001). Ontology development 101: A guide to creating your first ontology. Stanford Knowledge System Laboratory Technical Report KSL-01–05 and Stanford Medical Informatics Technical Report SMI-2001–0880.

OWL Web Ontology Language Overview. (2013). OWL Web Ontology Language Overview. http://www.w3.org/TR/owl-features.

Protégé. (2013). Protégé. http://protege.stanford.edu.

Top Braid Composer. (2013). Top Braid Composer. http://www.topbraidcomposer.com. Accessed 4 Jan 2014.

Zhao, J., Wroe, C., Goble, C., Stevens, R., Quan, D., & Greenwood, M. (2004). Using semantic web technologies for representing e-science provenance. Proceedings of the 3rd International Semantic Web Conference ISWC2004, Hiroshima, Japan, Springer LNCS 3298.

References

Chapter 10
Semantic Enrichment of Event Stream for Semantic Situation Awareness

Kia Teymourian and Adrian Paschke

10.1 Introduction

Complex event processing (CEP; Luckham 2001) is an enabling technology used to achieve actionable, situational knowledge from huge amounts of events in real time or almost close to real time. Detection, prediction, and mastery of complex events from event streams and event clouds are crucial to the competitiveness of networked businesses, the efficiency of the Internet of Services, and dynamic distributed infrastructures in manifold domains such as finance, logistics, automotive, telecommunication, and life sciences. The detection of complex events in organizations is used, for example, for optimized management of business processes and real-time decisions. One of the fundamental requirements of CEP systems is the time-critical processing of events. The efficient processing of events can be seen as fundamental quality-of-service (QoS) requirements of event processing systems (Chakravarthy and Jiang 2009). This real-time CEP behavior is considered as one of the main prerequisites for many highly relevant technology trends such as predictive business, real-time adaptive enterprise, or autonomic systems. CEP is now one of the fastest growing segments in (distributed) enterprise middleware software, with products provided by major software vendors and many start-up companies around the world. CEP has many use cases such as business activity monitoring (BAM), healthcare, fraud detection, smart offices/cities, logistics and cargo, information dissemination, event-driven adaptive systems, and supply chain management (Etzion and Niblett 2010).

Work performed while Kia Teymourian was affiliated with Freie Universität Berlin.

K. Teymourian (✉)
Department of Computer Science, Rice University, P.O. BOX 1892, TX 132, Houston, USA
e-mail: kia.teymourian@rice.edu

A. Paschke
AG Corporate Semantic Web, Institute for Computer Science,
Freie Universität Berlin, Berlin, Germany
e-mail: paschke@inf.fu-berlin.de

M. Workman (ed.), *Semantic Web*, DOI 10.1007/978-3-319-16658-2_10

Often complex events stem from complex factors and cannot be simply detected in the business activity workflows by using low-level syntactic definitions of event detection patterns. The permanent streams of low-level events in different business sectors need real-time semantic CEP (SCEP) that can profit from the large amounts of knowledge stored in the enterprise knowledge bases (KBs). The promises of the combination of event processing and semantic technologies are that these SCEP engines can use semantic background knowledge for defining more expressive event detection patterns, for understanding what is happening in terms of events and situations, and for knowing what rule-based actions they can invoke. The challenge lies in the combination of distributed real-time (big) data processing and semantic reasoning with background KBs.

This chapter first gives an overview on syntactic event processing techniques and then addresses the extension with semantic background knowledge, which leads to SCEP. It addresses the problem of optimized semantic querying, fusion and enrichment of real-time event streams with background knowledge, and the expressive reasoning with such enriched events in the higher layers of rule-based event processing and reaction functions.

10.2 Overview on Event Processing

Different event processing methods are developed to detect complex events, specified by different complex event expressions, from raw and primitive simple events. Different approaches are developed for event detection. A survey and requirements analysis about event processing methods are provided in Schmidt et al. (2008). A standardized reference architecture and typical event patterns are presented in Paschke et al. (2012a). A reference model and an overview on standards in CEP are given in Paschke et al. (2011).

Before we start with the description of different event processing approaches, it is important to have a better understanding of the main event processing concepts. The Event Processing Technical Society (EPTS) provided a reference architecture (Paschke et al. 2012a), reference model (Paschke et al. 2011), and glossary (Luckham and Schulte 2011) that defines the main concepts of event processing. "Event is anything that happens, or is contemplated as happening." *Event or event object (aka event message, event tuple) is an object that represents, encodes, or records an event, generally for the purpose of computer processing. A complex event is an event that summarizes, represents, or denotes a set of other events. Event pattern is a template containing event templates, relational operators, and variables. An event pattern can match sets of related events by replacing variables with values.*

Existing methods for event processing can be categorized into two main categories: rule-based approaches (Paschke and Boley 2009; Paschke and Kozlenkov 2009) and non-rule-based approaches (aka logic-based or non-logic-based approaches). Some of the non-rule-based event processing approaches are based on formalizations, such as finite-state automata (Gehani et al. 1992a, b), event graph (Chakravarthy and Mishra 1994; Paton and Diaz 1999), and Petri nets (Gatziu and Dittrich 1994). Various research prototypes and commercial products exist, which implement these event processing methods. In the following section, we briefly review the core event processing methods and some of the related implementations.

Finite-State Machine

Finite-state machine (FSM; aka finite-state automata) provides a simple computational model. A state machine can be in one of the finite number of states, and at a specific time point it is in only one of these states. In event processing, state machines can be used for event detection because the raw event stream can be considered as input sequence and the output of complex events can be considered as the writing of symbols. State machines are used in event processing in approaches coming from active databases such as Ordinary Differential Equation (ODE; Gehani et al. 1992a, b), COMPOSE (Gehani et al. 1993), Swiss Active Mechanism-Based Object-Oriented Database System (SAMOS; Gatziu and Dittrich 1992), and Active Directory Application Mode (ADAM; Diaz et al. 1991). In the following paragraph, we briefly describe recent event processing approaches (or CEP systems) that use finite-state machines.

Other event processing systems are SASE (Gyllstrom et al. 2006) and SASE+ (Diao et al. 2007) that are on nondeterministic finite automaton (NFA). SASE+ provides some of the typical event processing operators such as SEQUENCE, NEGATION, and sliding window operators. The event processing language of SASE+ provides Kleene closure (aka Kleene star or Kleene operator) over event streams that find multiple applications in radio frequency identification (RFID) data streams.

One further automaton-based system is Cayuga (Brenna et al. 2007) which is a research project at Cornell University Cayuga (http://www.cs.cornell.edu/bigreddata/cayuga) that provides a query language for the expression of complex event patterns called Cayuga Algebra. It also supports some special performance optimization such as indexing and garbage collector.

Esper (http://www.espertech.com/products) is an event processing engine that uses finite-state automata for event detection. Esper provides an event query language which has operators similar to Continuous Query Language (CQL), *SELECT, FROM,* and *WHERE,* and supports correlations and Structured Query Language (SQL)-like queries over event streams (Esper event processing language is a scripting language and supports some operators from CQL).

Esper also provides operators for the specification of event detection patterns and special operators to define the event stream consumption policy like *EVERY* operators that specify precisely how the event stream should be matched to the pattern. Event detection conditions can be specified over sliding time windows and can be used to trigger a reaction based on them. These actions are implemented in the Java programming language as Esper itself is written in Java.

Graph-Based Approaches

Chakravarthy et al. (1994) propose the use of the event detection graph (EDG) for the detection of composite events. In this approach, the event detection expression (complex event query) is built up as a graph pattern like a tree structure, so that different event types are defined as the leaves of a tree and nodes are the event operation algebras such as AND, OR, SEQ, and NOT. The events are then streamed into this structure, and in each node one of the rules are executed on the event stream. Gatziu and Dittrich (1994) propose the use of colored Petri nets (Jensen 1996) in the SAMOS system for the detection of composite events.

Rule-Based Approaches

A general introduction on rules and logic programming on the Web is given in Paschke (2009). Rule-based event processing is addressed in Paschke and Boley (2009) and Paschke and Kozlenkov (2009). Several industrial event processing products use rule engines for event processing without permanent storage of events (the storage of historical event data is only optional for other purposes). Rule-based event processing engines can process events in real time while keeping the complete rule set (including facts) in the main processing memory. However, without any further optimization of their semantic reasoning capabilities, these approaches cannot achieve high scalability and high performance, when they have to process huge amounts of semantic background knowledge (similar to static reference data) for semantic event detection. The main problem here is that they have to keep the whole KB in the main processing memory. This will be infeasible when the background knowledge is very large, for example, all the background knowledge about companies traded on the stock exchange market worldwide.

One of the rule-based approaches is introduced in Paschke (2007) which proposes a homogeneous reaction rule language for CEP. It is an approach combining event and action processing, formalization of reaction rules in combination with other rule types such as derivation rules, integrity constraints, and transactional knowledge.

In addition, several event processing languages have been proposed such as Snoop (Agrawal et al. 2008), Cayuga Event Language (Brenna et al. 2007; Demers et al. 2006), SASE (Gyllstrom et al. 2006), and XChangeEQ (Bry and Eckert 2007). Some of the commercial CEP products are TIBCO *BusinessEvents* (http://www.tibco.com/products/event-processing), Microsoft *StreamInsight* (http://blogs.msdn.com/b/streaminsight/), and Sybase CEP (http://www.sybase.com/products/). *Reaction RuleML* (Paschke and Boley 2009; Paschke et al. 2012b; Paschke 2014) is a platform-independent rule interchange standard for reaction rules and rule-based CEP.

Some of these CEP systems can integrate and access external static or reference data sources. But these systems do not provide any (semantic) inferences on external KBs and do not consider reasoning on relationships of events to other non-event concepts.

Prova (Prolog + Java http://www.prova.ws/) (Kozlenkov et al. 2006) is a rule language and a rule engine. Prova provides an open-source rule language whose design is based on reactive messaging, and a combination of imperative, declarative, and functional programming in serial Horn rules. One of the important design principles in Prova is reactive messaging that allows the organization of several Prova rule processing engines into a network of communicating agents. A Prova agent is a rule base that is able to send messages to other Prova agents by using primitive message-passing primitives. The reactive messaging functionality in Prova makes it possible to build workflows based on communicating distributed Prova agents. In both sending and receiving message primitives, five different parameters are sent between the agents. The parameters are: a conversation ID of the message (XID); name of the message-passing protocol (Protocol); Destination (on sending) or Sender (on receiving); the message type broadly characterizing the meaning of the message (Performative); and the message itself, a list containing the actual content of the message (Payload). The event processing functionality in Prova uses reactive messaging and rule-based workflows. An event processing graph can be mapped to a reactive message-passing workflow into Prova. Prova uses metadata attributes like (AND, OR) to build the workflows and thus event algebra operations.

One of the recent rule-based systems is Event-Driven Transaction Logic Inference System (ETALIS; Anicic et al. 2011a). ETALIS is a rule-based stream reasoning and CEP. ETALIS is implemented in Prolog, and it uses the underlying Prolog-inference engine for event processing. A major distinction between ETALIS and Prova is that ETALIS is a meta-program implemented on top of a Prolog system with only one global KB in which every piece of knowledge, such as incoming events, is globally applied. Prova instead allows for local modularization of the KB and scoped event processing states within complex event computations and event message-based conversations. This leads to a branching logic with local state transitions as it is common, for example, in workflow systems and distributed parallel processing.

RETE Algorithm in Event Processing

The RETE algorithm provided by Forgy (1979) is a logical matching algorithm for matching data tuples (or facts) to the rules in a pattern-matching system. The RETE algorithm is based on the forward chaining inference with facts and rules (Paschke and Boley 2009).

The RETE algorithm builds directed acyclic graphs as a high-level representation of the given rule sets, which are generated in run-time and includes objects such as nodes of the network. All of the operations on data tuples such as relational query processor, performing projections, selections, and joins are executed on the network of objects.

In RETE, facts are loaded into the memory, and the rule engine creates working memory elements (WMEs) for each of the facts. Each WME includes a set of n-tuples. The RETE network is divided into two subnetworks: *Alpha* and *Beta* networks. The left side of the network is called Alpha network and includes the network part that is responsible for selecting individual WMEs based on conditional matching to WME attributes. Conditional matching may include several tests within the network for the testing of several attributes. The right side of the network is called Beta network and performs the join operations on WMEs. The join operations are down on the interim results from two other nodes, and the join result is stored in a beta memory node. The results from the join nodes are processed in the final stages by terminal nodes, so that they produce the final agenda of the rule set given to the rule engine.

The RETE algorithm has been used as an event processing engine, for example, Walzer et al. (2007) because the inserted facts can be considered as event messages that arrive into the system as data stream, and the event detection pattern is the given rule set. RETE has also been used in commercial event processing products such as TIBCO Business Events (http://www.tibco.com/products/event-processing/) and Drools Fusion (http://www.drools.org/).

Miranker (1987) describes the advantages and disadvantages of the RETE algorithm. The main advantage of RETE is that the large amount of interim results and states stored in the memory nodes minimizes the number of comparisons of two WMEs, and the stored similar results are reused and shared with other WME tests. By sharing the structural similarity in rules, RETE can speed up the rule-matching process.

The drawbacks of the RETE algorithm are twofold: the primary disadvantage of RETE are WME updates, for example, removal of one of the WMEs causes a restart of the whole calculation and repetition of the entire sequence process upon its addition to the network. The second disadvantage is the storage of states in memory nodes which is highly memory intensive and may be combinatorial expensive. Thus, sharing the network structure might not be realizable in a parallel environment due to the communication costs. The performance of the original RETE algorithm is improved by further optimization approaches such as RETE II (http://www.pst.com/reteII.html).

Storage-Based Event Processing

The basic and more naive approach might be storing incoming event data in a database and steadily querying and pulling the database. The main disadvantage of this approach is that processing is possible only after storage and that the database is pulled with each new incoming event. This approach can work for use cases which do not have high event throughput and huge amount of background knowledge to process. The advantage of this approach is that a complete reasoning on the whole knowledge inventory is possible. Scalability and real-time processing are the problems of this approach which makes it impossible to use it for time-sensitive use cases such as algorithmic trading or fraud-detection systems. The usage of distributed databases can improve scalability, but it can negatively affect real-time processing latency.

Glombiewski et al. (2013) showed that a standard database can be used for event processing by using data access technologies such as Java database connectivity (JDBC; http://www.oracle.com/technetwork/java/javase/jdbc/index.html; JDBC is a Java-based data access technology). Their experiments showed that such event processing systems can handle small- and medium-sized event processing workloads.

In recent years, emerging technologies in computer hardware technology make it possible for main memory hardware to be cheaper than ever before (Stonebraker et al. 2007). Schapranow and Plattner (2013) describe blade servers as providing roughly 500 Gb of main memory.

The idea behind in-memory databases is to keep the complete database in the main memory so that database management systems (DBMS) can gain better performance than conventional disk-based database techniques (Willhalm et al. 2009). Many in-memory database prototypes or commercial systems have been developed, such as Hyper (Kemper et al. 2012), MonetDB (Manegold et al. 2009), H-Store (Stonebraker et al. 2007), and HANA (Willhalm et al. 2009).

Also, several data stream processing systems have been proposed so far such as Telegraph (Chandrasekaran et al. 2003) and Stream (Group 2003). Data stream processing systems aim at handling continuous database queries over high-throughput data streams. These systems are similar to event processing systems and have similar properties (Chakravarthy and Jiang 2009).

Data Stream Processing Systems

In the following, we review some of the most relevant data stream processing system. STREAM, the Stanford Stream Data Manager (Group 2003), is a data stream management system (DSMS) that uses a declarative rule language called CQL. It is an extension of SQL for querying streams over sliding windows which are either time- or tuple-based data windows.

The semantics of CQL is composed of three main parts:

- A relational query language that includes relation-to-relation operators
- A window specification language that can specify data windows using stream-to-relation operators
- A set of relation-to-stream operators for the specification of interactions from the relations to the stream

Kraemer and Seeger (2004) propose an infrastructure called Public Infrastructure for Processing and Exploring Streams (*PIPES*) for providing flexible and extensible DSMS building blocks. *PIPES* covers the operational semantics and functionality of the CQL. The language provides options for query optimization which is based on approaches adopted from the temporal database community. Query optimizations enable applications of temporal transformation rules within the context of streams.

Chandrasekaran et al. (2003) propose another DSMS called TelegraphCQ which uses a continuous query language based on stream-only approach. TelegraphCQ includes entities named *Eddies* that are used for routing tuples through a network of query modules. Continuous queries can be specified over sliding time-based windows. Telegraph is implemented based on the PostgreSQL (http://www.postgresql.org/) database system and modifies it for the processing of streaming data.

Gigascope is another DBMS provided by Cranor et al. (2003). Gigascope is specially developed for network applications such as traffic analysis, intrusion detection, router configuration analysis, network research, and network monitoring. Gigascope uses a query language called geographic structured query language (GSQL) which is a pure stream query language following the SQL syntax, and it supports *join, selection, aggregation,* and an additional stream merge operator.

Another, DBMS is Aurora (Abadi et al. 2003) whose aim is to process streams of data coming from various sources. Aurora provides a stream query algebra called SQuAl that includes seven primitive operations, for example, to select, filter, and aggregate.

10.3 Semantic Complex Event Processing

Semantic models of events can improve event processing quality by using event metadata in combination with linked data, ontologies, and rules (semantic KBs). In recent years, huge amounts of public background knowledge became available on the Web as Linked Open Data (LOD; http://linkeddata.org) and ontology and rule repositories. (The term "semantic complex event processing" can be confusing. It can be confused with the formal semantics of event detection rules. With the term "semantic," we refer to the usage of ontological background knowledge about events and other resources in the application area.) Organizations can access these background data, filter, enrich, and integrate them with the organization's internal knowledge to build high-quality and trustworthy background knowledge about their specific business application domain.

The combination of event processing and semantic knowledge representations can lead to novel semantic-rich event processing engines. These intelligent event processing engines can understand what happens in terms of events, can (process) state and know what reactions and processes they can invoke, and furthermore, what new actions/events it can signal. The identification of critical events and situations requires processing vast amounts of data and metadata within and outside the systems.

One of the promising areas of application is *SCEP* (or synonym knowledge-based CEP; Teymourian and Paschke 2009a, b, 2014; Teymourian et al. 2012), which is the main subject of this chapter. Complex events are events which are not explicitly defined before event processing, but a system user can describe them in a high level and abstract event pattern language based on background knowledge of the target application domain.

In most of the CEP use cases, the usage of background knowledge can improve the quality of event processing. The CEP system can profit from using semantics of events in combination with existing knowledge about the target application domain. The benefits of the fusion of ontological background knowledge with event processing are the same as using metadata in data-driven applications, such as extracting more relevant and useful complex events and flexibility that enables interoperability and extensibility.

Complex events, patterns, and reactions can be precisely expressed and can be directly translated into business operations. New business changes can be integrated into CEP systems in a fraction of the time, while complex event patterns are independent of actual business and are defined based on abstract business strategies.

Some of the motivating examples for SCEP use cases are presented below to illustrate the benefits of using background knowledge in the domain of CEP. One example of complex events can be found in eHealth. A patient in a critical, life-threatening situation can be seen as one such complex event. Patient can be monitored at hospitals (Wienhofen and Toussaint 2010) or at home using different body-sensing devices, so-called RFID readers. These technologies make it possible to capture huge amount of streaming data about the vitality signs of patients. The exact diagnosis whether a patient is in a critical situation can only be done by the medical staff. However, the medical staff can be informed about potential critical situations, so that they can react faster. A system can support these decisions and detect complex events based on the event stream of sensor data which are emitted from different body sensors and other tracking devices.

A classic example in the area of event processing and data stream processing is the real-time analysis of stock market events generated by the stock market exchange. The decision to buy or sell company shares depends on real-time information and background knowledge about different companies. Stock brokers develop their negotiation strategies based on chunks of information that they gather from their resources. Their strategy may depend not only on stock market prices and the volumes handled but also more on a combination of the companies' attributes and relationships (e.g., to other organizations or to their customers) in a specific temporal context. A buy or sell decision can be based on market status, company products, company staff, and all of the situational attributes of companies. Most of today's monitoring and automated handling systems work on the basis of the syntactic processing of price/volume event stream, but not based on background knowledge about the companies. For example, a stock market broker might be interested in the shares of *companies* which have *production facilities in Europe,* produce products from *iron,* have more than 10,000 employees, are at the moment in *restructuring phase,* and their price/volume *has been increasing continuously* in the *past 5 min.*

They can only deal with it in the case of increasing/decreasing of prices/volume around predefined levels. The brokers might be able to describe it in a higher level and more abstract language, which can be processed by intelligent processing systems.

Issues and Challenges

The processing approach of current event processing engines often relies on the processing of simple event signals, and events are merely implementation issues. The existing approaches provide only inadequate expressiveness to describe background knowledge about events and other resources in background knowledge which are related to event processing. They do not provide adequate description methods for complex decisions, behavioral logics including expressive situations, pre- and post-conditions, complex transactional (re-) actions, and work-flow-like executions.

All of these are needed to declaratively represent real-world domain problems on a higher level of abstraction. The identification of critical complex events requires processing vast amounts of data and metadata within and outside the event processing systems. For some of application scenarios, an intelligent event processing engine is required which can understand, what happens in terms of events and can (process) state and know what reactions and processes it can invoke, and which new events it can signal.

The fusion of event processing approaches and knowledge representation methods can lead to more knowledgeable event processing systems. However, the existing knowledge representations and inference techniques are not directly suitable for event processing applications because of the different requirements of event processing applications such as real-time processing of events.

Knowledge representation for events goes beyond defining event types and their hierarchies. The relationships of events to other nonevent concepts in the application domain are part of specifying complex events. Background knowledge for event processing can be integrated from the domain- and application-specific ontologies for events, processes, states, actions, and other concepts. Specific domain, task, and application ontologies need to be dynamically connected and integrated into the respective event processing applications, which also leads to a modular integration approach for these ontologies. Capturing domain-specific complex events and generating complex reactions based on them are fundamental challenges.

The main problem is the utilization of background knowledge about events and other nonevent concepts and objects for a SCEP. This integration enhances the expressiveness of event processing semantics and makes the event processing systems more flexible. More precisely, we consider one or more event streams which include highly frequent events of different types and one or more KBs, which include background knowledge about the events and other related resources, in the target application domain. Because background knowledge can be huge, we consider that the metadata has to be stored in external KBs and cannot be handled completely in the main memory of a single event processing engine.

Lack of Knowledge Representation Methods Event processing needs a knowledge representation formalism. The current event processing systems do not provide any semantic knowledge representation methods for events, and there is no precise logical semantics about other related concepts and objects in the target domain.

A formal specification builds a stable foundation which is needed for description and reasoning. A formal specification also avoids semantic ambiguity. Event processing needs as its basis a formalization and specification which can describe simple events, complex event patterns, situations, pre- and post-conditions, (re-) actions, and other related concepts. Definition of events by logic is not addressed in state-of-the-art syntactic event processing solutions. Events have special characteristics and attributes which differentiate them from other objects. Events can

happen. They can be considered as entities which unfold over time and exist only over a specific time window. The questions related to this problem are as follows:
How should raw events and complex events (event patterns) be represented based on relations of resources in background knowledge? How should knowledge about events and event patterns be represented?
Which extensions are required, so that a query language for detecting the complex events based on knowledge graph patterns and event operation algebra can be specified?
How should ontologies about events, actions, states, situations, and other related concepts be created and managed from the ontology engineering perspective, when we have to take into account the restrictions of event processing?

We specify complex event patterns based on the relations existing in background knowledge to improve the expressiveness and flexibility (generalization) of the CEP patterns.

Limitation of Processing Methods for Fusion of Events and Background Knowledge The existing event processing methods can only detect and process events based on their syntax and incoming (temporal) sequence. The expressiveness of event processing decreases when the system uses ontological knowledge about the events and their environment and combines it with detection and reaction rules.

One of the obvious questions raised in the course of working on event processing systems is the complexity to which event processing engines can detect and process complex events. This question is difficult to answer because the capability of processing complex events when compared to each other depends on various parameters. These parameters can only be defined for specific use cases on a case-by-case basis. Because of this, the existing evaluations of event processing engines have not been able to answer this question properly. The expressiveness of event processing increases when the system uses ontological knowledge about the events and their environment and combines it with reaction rules.

Event processing approaches should be extended, so that they can include ontological semantics of events, processes, states, actions, and other concepts into event processing without affecting the scalability and real-time processing. The research questions regarding this topic are:
Is the use of additional semantic background knowledge in event detection queries a limitation for real-time event processing or scalability of the event processing system?
To what extent can we optimize the trade-off between the complexity of reasoning with expressive background knowledge (e.g., description logics reasoning) and real-time event processing? To what extent is the fusion of background knowledge with the event data stream a bottleneck for real-time event processing? Is it possible to avoid this bottleneck by using query optimization approaches and distribution on networks of processing nodes?
What are the assets and drawbacks of event enrichment approaches when they should handle huge amounts of background data and run computational intensive reasoning on external KBs?
Is it possible to preprocess detection queries to optimize their event detection query plans and select the most optimal one based on execution costs?

Applications of Knowledge-Based Event Processing

CEP has many different use cases which profit from the real-time event detection and reactions to detected complex events. Some of these use cases are BAM, fraud detection, smart offices/cities, logistics and cargo, information dissemination, event-driven adaptive systems, supply chain management, and intelligent systems in health care.

The extension of CEP to knowledge-based event processing enables CEP systems to profit from domain background knowledge and enables a more knowledge-intensive event processing. In all of the use cases mentioned, knowledge-based CEP can improve the quality of event processing and profit from using event semantics in combination with existing knowledge about other concepts in the target application domain. The benefits of adding ontological background knowledge to the area of event processing is the same as data processing.

The use cases of knowledge-based/semantic CEP have different requirements for the processing of events and integration of knowledge from external KBs. We categorize these requirements in the following points:

Real-Time Processing of Events Different use cases might be different with respect to meeting detection deadlines. Some of them might have absolute hard deadlines for the detection of events, while some others can tolerate the missing of deadlines, or only a predictable number of deadlines.

Scalability of CEP System The ability to process high-throughput event streams in addition to the ability to fuse large-size external KBs.

Reasoning on Background Knowledge The level of required reasoning on background knowledge can be different from use case to use cases.

Expressiveness of Event Detection Queries The complexity of event detection rules can be different.

Elasticity of SCEP System The requirement to process data with highly variable sizes/dimensions on demand and be able to optimize processing costs.

Requirement for Guarantee Detection In some of the use cases, it is required that the system detects all complex events and provides a granite for the detection of complex events that have happened. In some other use cases, it is accepted if the system provides an approximation of the complex events and does not detect all of them, that is, detect some of the complex events, but be able to detect these events in real time.

Example Use Case: High-Level Market Monitoring

In today's economy, companies are highly interconnected and dependent on each other. For example, companies require raw material, suppliers, and financial credit.

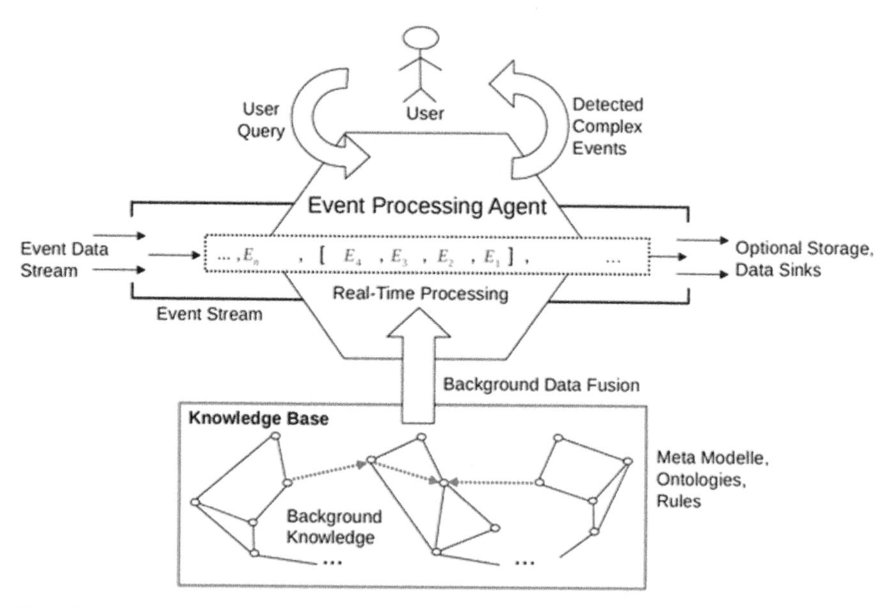

Fig. 10.1 Example of relations on stock market events and companies background knowledge

Businesses are also dependent on laws legislated by politicians. We can see some kind of domino effects among companies which impact on their businesses and the price of their stock market shares. In the case that the business of one of the companies changes, the business of the other companies might be affected as well.

The question in this use case is if it is possible to detect complex events according to companies' relations, so that decision-makers can react to them in a timely manner. The target is not to predict the start of the domino effect, but to notify as soon as the first stones fall.

A complex event detection pattern is shown in Fig. 10.1 which is defined based on the background knowledge about companies and company dependencies. It shows stock market events and insight into the related background knowledge about the companies. We can see that companies have some business dependencies on each other, company Y1 produces raw material M1, the business of another company Y2 depends on this raw material for its production and might have big trouble if they cannot supply the material. A third company X3 finances company Y2 and might have some financial problems if company Y2 gets material shortage trouble. A stock market broker might be interested in this dependency chain and can define a complex event detection pattern for this particular complex event without even knowing what these companies are. He might be interested to know when the stock prices of these three companies start falling. The aim of this event detection is not to predict stock market prices, but to be informed when the prices of these three companies fall.

```
{ .., {(Name, ``GM") (Price, 20.24) (Volume, 8,835)},
      {(Name, ``SAP") (Price, 48.71) (Volume, 8,703)},
      {(Name, ``MSFT") (Price, 24.88) (Volume, 46,829)},  ... }
```

Suppose Mr. Smith is a stock broker and has access to a stock exchange event stream, like those listed in the above example. He is interested in particular kinds of stocks and would like to be informed when there are some interesting stocks available for sale. His particular interest or his particular stock handling strategy can be described in a high-level language which describes the interest using background knowledge about companies.

Mr. Smith would like to detect complex events using the abstract query listed in the following:

```
Select stocks of companies which
        have production facilities in Europe,
        produce products from Iron,
        have more than 10,000 employees,
        are at the moment in restructuring phase and
        their price/volume have been increasing in the past 5 minutes.
```

As we can see, the above query cannot be processed without having background knowledge to define the concepts in this query. Mr. Smith needs an intelligent system which can use background knowledge about companies like those listed in the following listing. This background knowledge should be integrated and processed together with the event data stream in a real-time manner, so that interesting complex events can be timely detected.

{ (OPEL, belongsTO, GM), (OPEL, isA, automobilCompany),

(automobilCompany, build, Cars), (Cars, areFrom, Iron),

(OPEL, hatProductionFacilitiesIn, Germany),

(Germany, isIn, Europe),

(OPEL, isA, MajorCorporation),

(MajorCorporation, have, over10,000employees),

(OPEL, isIn, reconstructionPhase), ... }

We can also assume that Mr. Smith works for a company and may need to share this KB with other brokers. Each of these brokers may be able to gather new information about companies and update this KB, for example, the Opel Company is not in restructuring phase, or a company has a new chief executive officer.

Raw Event Source Live stream of stock market rates from different stock markets.

Background Knowledge Resource Background data about stock market companies, e.g., extracted from LOD KBs and integrated into the corporation's internal knowledge.

Complex Event Type Events are defined based on company relations and the relations of other resources to other companies.

Derived Knowledge Extraction of further company relations.

Real-Time Factor In the case that some of the detection deadlines have been missed, financial disadvantages are expected. This can be considered as a soft real-time system in which the number of times that the system misses the deadline is predictable.

Approaches for Knowledge-Based Event Processing

The area of (semantic) CEP is an emerging research area and has been increasingly gaining attention in the past years. Some of the recent efforts in the area of event processing are focused on the usage of ontological background knowledge to improve event processing expressiveness.

In the following section, we briefly describe these approaches including approaches for semantic event processing and stream reasoning. We discuss the differences between these approaches and an approach for multistep event enrichment and detection.

One of the initial efforts on semantic event processing is Semantic Toronto Publish/Subscribe System (*SToPSS*) presented by Petrovic et al. (2003). It extends the conventional event processing in publish/subscribe system with some semantic capabilities. Taxonomies are created that provide a conceptualization of hierarchical relationships between the concepts and the synonyms of the concepts. These taxonomies are then used in the matching processes of events to user-specified event patterns. By using the taxonomies, the event patterns are expanded to further matches of streaming events. The fundamentals of the semantic enrichment of event streams are proposed, which have been further extended by using external KBs to include complex ontologies with different level of expressiveness.

Zhou et al. (2012) provide another system called *SCEPter* which can enrich event object with additional knowledge extracted by inferences from ontologies. Annotations are used to filter or group different event objects by using their event types. *SCEPter* is able to extract single event instances based on their background knowledge. The approach provided does not target the detection of groups of events based on their relation chain in background knowledge.

Stream Reasoning and Resource Description Framework Processing Systems

Valle et al. (2009) proposed stream reasoning, a novel approach for the integration of data stream processing, the Semantic Web, and logical reasoning systems. Stream reasoning technologies should provide foundations, methods, and tools for the integration of streaming world with the more static web of data.

Sequeda and Corcho (2009) propose the concepts and visions of Linked Stream Data (LSD) which applies the Linked Data (LOD, http://linkeddata.org) principles to streaming data. LSD should allow publication of data stream in combination with Linked Data Web.

Several stream reasoning systems and approaches have been developed, such as CQELS (Le-Phuoc et al. 2011), streaming SPARQL (Simple Protocol And RDF Query Language (SPARQL) is a W3C recommendation and a query language for RDF) (Bolles et al. 2008), and SPARQLStream (Calbimonte et al. 2010). In most existing stream reasoning approaches, the streaming data are in the Resource Description Framework (RDF) data format, and the stream reasoner has the task of providing reasoning on a sliding window of RDF data.

Some stream reasoning languages and processing approaches have also been proposed. Barbieri et al. (2010) propose continuous SPARQL (C-SPARQL) and event processing SPARQL (EP-SPARQL) (Anicic et al. 2011a) as a language for continuous query processing and stream reasoning.

ETALIS provides two event processing languages: ETALIS Language for Events (ELE) and EP-SPARQL. ELE provides features such as classic event operators and count-based sliding windows. EP-SPARQL (Anicic et al. 2011b) is a language for complex events and stream reasoning. The formal semantics of EP-

SPARQL is along the same lines as SPARQL (RDF Query Language SPARQL http://www.w3.org/TR/rdf-sparql-query/). EP-SPARQL can be used in ETALIS for reasoning on RDF triple stream (event stream can be mapped to RDF stream). EP-SPARQL provides additional operators such as SEQ, EQUALS, OPTIONALSEQ, and EQUALSOPTIONAL which can be used to define complex pattern to detect complex events in the stream of RDF data. EP-SPARQL can be used with the ETALIS event processing engine.

Le-Phuoc et al. (2011) present Continuous Query Evaluation over Linked Streams (CQELS), a native and adaptive query processor for unified query processing over RDF stream and linked RDF data. CQELS processing engine can optimize query processing by continuously reordering of operators according to some heuristics about the streaming data. Furthermore, the external disk access is improved by data encoding and caching of intermediate query results.

The CQEL query language (https://code.google.com/p/cqels/wiki/CQELS_language) provides window operators that allow specification of sliding window over RDF Streams.

Linked Stream Benchmark (LSBench) (Le-Phuoc et al. 2011) benchmarks CQELS against C-SPARQL and JTALIS (the Java wrapper for ETALIS). SRBench (Zhang et al. 2012) is a general purpose benchmark for streaming RDF engines. The benchmark uses weather sensor data, since they are inherently stream based and time bound, to model a realistic use case.

Although some of the existing approaches either directly work on RDF streams or allow querying external data sources, they do not address the detection of semantic enriched events on the basis of huge amounts of existing background knowledge. Most of the systems are designed for main memory event processing with pure syntactic event pattern matching or simple RDF pattern matching, without any further expressive semantic support.

It is possible to address the SCEP, the problem as a hybrid approach—expressive reasoning on background knowledge to be used in high-performance, real-time event processing. None of the existing systems is optimized for such hybrid event processing over huge amounts of semantic background knowledge (low frequently changing) and high-throughput event streams. In the following, we discuss the trade-off between high expressiveness of used background knowledge and high levels of computational complexity, which has a direct effect on the efficiency and scalability needed in real-time event processing.

10.4 Fusion of Event Stream and Knowledge

For the realization of SCEP, the required knowledge can be considered as an external KB which can manage background knowledge (e.g., conceptual and assertional, TBox, and ABox of an ontology) about events and other nonevent resources. The use of an external KB in event processing allows detecting events based on reasoning on a type hierarchy, temporal/spatial relations, or relations to other concepts/objects in the application domain.

An example for the connections of events to relevant concepts in background knowledge can be the relationship of a stock market event (e.g., a significant price change) to the products or services of that company (in a stock market monitoring use case).

A complex event query can be defined based on the terminology of ontologies and schema of event data streams. This allows users to define queries on a more abstract level, based on relationships in the background knowledge, and not only based on syntactic matching of the data values of events. A specified event query should be continuously evaluated against data coming from both sides, the event stream and the external KB.

As described before, most existing approaches for event processing use rule engines or finite-state automaton for the processing of events without permanent storage or indexing of event data. The event stream can flow through the system without any necessary storage. The storage of historical event data is only optional for purposes other than event detection.

Rule-based event processing engines can process events in real time, because they can handle whole facts and rules in the main processing memory. However, these approaches cannot achieve high scalability or high performance, when they have to process huge amounts of domain background knowledge for event detection. The challenging problem is that rule engines have to keep the whole KB in the main processing memory space. However, for very large KBs it is impossible to keep the completely inferred background knowledge for event processing in memory, for example, using background knowledge about the companies traded on the stock exchange markets worldwide.

To integrate and aggregate domain background knowledge with incoming event stream and to timely process of integrated knowledge, a high-scalable and high-performance processing approach is required. Particularly, inferencing on huge amounts of background knowledge can be a time and computation-intensive process. An implementation of SCEP needs to provide acceptable quality-of-service metrics such as low latency, high throughput, and high scalability. On the other hand, it also needs to provide expressive reasoning on background knowledge and derive events based on the inference on background knowledge. We can see that there is a trade-off between all of these metrics which should be optimized for the target use case. The abstract view of the proposed SCEP approach is shown in Fig. 10.2.

It illustrates the usage of an external KB which includes ontologies and rules, and an incoming event stream which comes from event producers, for example, sensors or event adapters. The SCEP engine should process the event stream based on semantic relations of event stored in the external KB. The KB includes a TBox (assertions on concepts) and the primitive event objects as ABox (assertions on individuals) which are used for semantic reasoning on events. As mentioned, in SCEP a part of the knowledge about events might be the relatively static knowledge about pre-defined event classes, that is, the event types in an event ontology, while the other part are real-time data streams. The system has to combine these knowledge references and infer new knowledge. The results of CEP are the detected complex

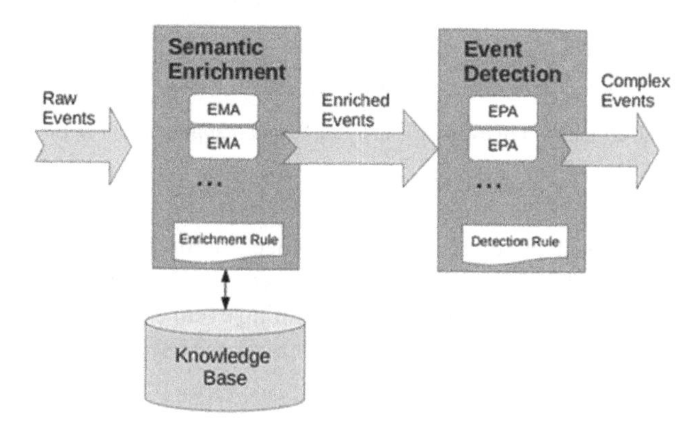

Fig. 10.2 Abstract overview of the semantic event processing using external KB. *EMA* event mapping agents, *EPA* event processing agents

events that are notified to the users. The CEP engine can also identify critical complex actions which might be triggered after these events happen and are detected.

A straightforward approach for SCEP might be a storage-based approach, for example, one can store all of the background knowledge on an external KB and start polling the KB on every incoming single event. The results from the KB can then be processed with an event data stream. This approach typically suffers from scalability and performance problems when the throughput of the event stream is high or the size of the background knowledge is huge, or even when expressive reasoning should be done on the KB. The main disadvantage of this approach is that processing is only possible after storing/indexing the data and the databases are polled with each new incoming event. This implementation might be applicable for use cases which do not have high event throughput or huge amount of background knowledge to process. The advantage of this approach is that a complete reasoning on the whole KB inventory is possible. Scalability and real-time processing are the problems of this approach which makes it impossible to use for time-sensitive use cases.

One of the approaches for the fusion of events and background knowledge is the enrichment of the event stream prior to event processing so that event processing agents can process events based on background knowledge relations. This chapter provides a description of concepts for the semantic enrichment of event streams. We describe the different components of the event enrichment process and illustrate possible architectures to achieve the required scalability and throughput. We analyze the different costs of event enrichment and describe the most important impact factors.

For improvement of scalability and throughput performance of knowledge-based event processing, we propose the Semantic Enrichment of Event Streams (SEES; Teymourian and Paschke 2014) approach, which is an approach for the enrichment of events with ontological background knowledge prior to a syntactic event processing step. The event stream is enriched with newly derived events or newly added attribute/values to each event instance. The events derived are generated from raw events and are only generated for internal usage.

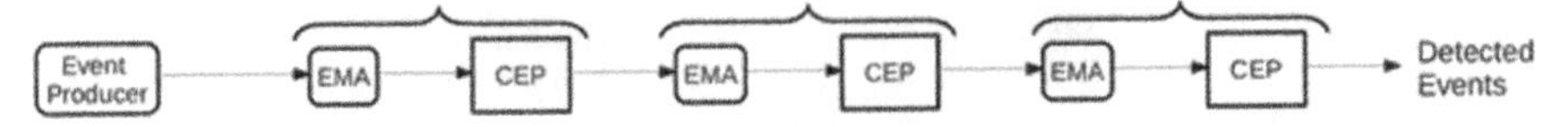

Fig. 10.3 Semantic enrichment of event streams. *EMA* event mapping agents, *CEP* complex event processing

The SEES approach extends existing approaches for event enrichment with further concepts for the detection of events based on their complex background knowledge and with providing concepts for planning event enrichment process. The planning of event enrichment and detection steps can be used to optimize the costs of event enrichment.

Semantic Enrichment

SEES is about the enrichment of events with background knowledge prior to complex event detection with new derived event attributes. In an event processing network (EPN), several event mapping agents (EMA) have the task of generating derived events by querying external KB and executing reasoning on existing knowledge. Mapping agents can be replicated to achieve better scalability. In the following processing step, the enriched event stream can be monitored by multiple event processing agents. EMAs can be replicated and deployed in parallel to achieve efficient scalability with respect to throughput.

An *EMA* is a software system that has the task of receiving events and generating newly derived events by querying external KB and mapping incoming events to new events. Derived events are completely new events or are received raw events with updated attributes (adding new fields or removing old ones).

In order to enrich the event stream, new events can be derived from raw event instances. Such derived events can contain attributes that are inferred from external KBs. Figure 10.3 illustrates the process of semantic enrichment of an event stream. As discussed before, the raw event stream is enriched by one or many EMA resulting in an enriched outbound event stream processed by a set of event processing agents (EPAs) in order to detect complex events.

A parallel setup of event enrichment in an EPN is possible. The main event stream is split into several sub-streams. The task of event enrichment can be distributed and shuffled into several EMAs. Each EMA uses a shared or dedicated KB. Task scheduling can be realized by using, for example, simple round-robin scheduling or load distribution heuristics on the processing capacity and load of each EMA. The parallel setup of the event enrichment approach is similar to the concept of map reduce (Dean and Ghemawat 2008), incoming events are mapped to new events, and then a CEP engine reduces the events to detected complex events.

Figure 10.4 shows an example of the SEES approach for multistep knowledge acquisition and event detection. Each SCEP node includes an EMA and EPA engine, so that it can enrich and detect events. At the initial processing step (the starting

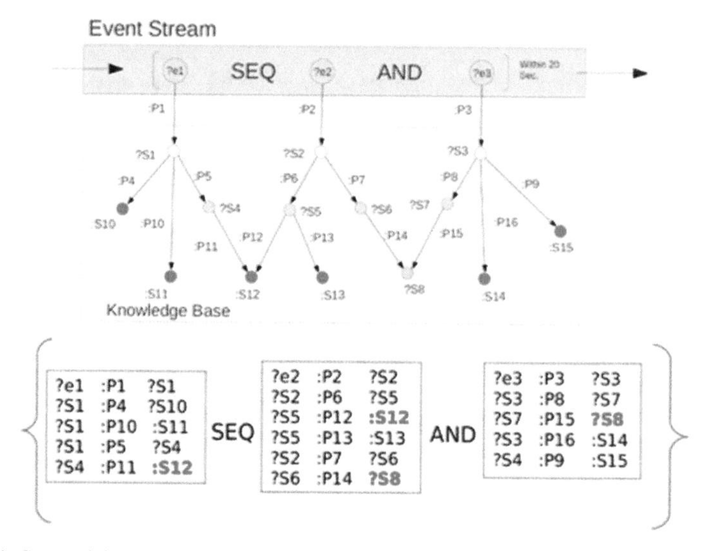

Fig. 10.4 Sequential setting of multistep greedy knowledge acquisition for SCEP

point of event processing), the raw event stream flows through the system. In the first step, an EMA node enriches the event stream with knowledge acquired by using a subgraph of the query graph pattern, and the EPA can filter out a subset of event instances depending on the knowledge used. We observed that a partial enrichment of events can be used to filter out some of the events to avoid unnecessary enrichment of all raw events.

Figure 10.3 illustrates a sequential setting of multistep event enrichment and detection. In each step, the input event stream is enriched, and the event detection process is executed on the enriched event stream. The stepwise processing of events is useful for avoiding the full enrichment of the complete event stream; in each step, a module of background knowledge can be enriched to event stream and used for event detection. In this way, the system can optimize the cost for event enrichment. A combination of both parallel and serial approaches is also possible, so that in each parallel enrichment process a sequential enrichment process is organized to optimize enrichment costs.

Plan-Based Semantic Enrichment

The process of semantic enrichment can be optimized to reduce the cost of event enrichment and increase throughput of event processing by reducing the amount of raw event enrichment tasks. We propose an approach for the optimization of knowledge-based event detection by using a technique for multistep and greedy knowledge acquisition and event detection. In the SEES approach, we use sequential setting and several steps for event stream enrichment and detection of complex

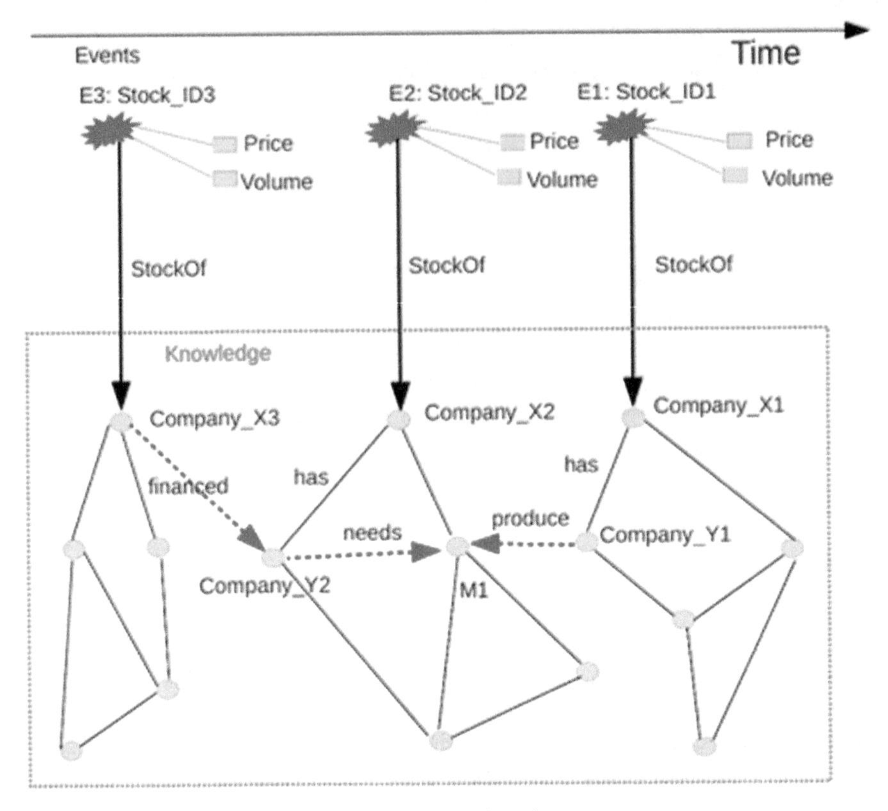

Fig. 10.5 An example of a knowledge-based event detection pattern

events from the enriched stream. In each step, a part of the knowledge is used to enrich the events. The event detection engine then can filter out some of the raw events based on enriched knowledge, so that only relevant raw events are forwarded to the next step. By using this approach, we can avoid the unnecessary full enrichment of all raw event instances.

The trade-off between knowledge acquisition costs (computation load on external KB and result transmission) and event processing latency is an important factor for planning the execution of event enrichment and detection. The aim is to discover a low-cost event detection plan while meeting user-specified latency expectations, so that we can reduce the polling load on the external KB. One of the important constrains for generating plans is user-specified latency expectation. We search for a plan which can meet this expectation and generate an acceptable load on the KB side. In the final step, we have to match the whole user query on the event stream; we look only for good filters to pre-filter the event stream so that we can reduce costs.

Figure 10.5 shows an example of knowledge-based event detection pattern. Such kind of user event detection pattern (user query) can be preprocessed and separated into several subqueries. In the SEES approach, we generate a plan for stepwise processing of the generated subqueries, so that we can pre-filter the raw event stream to reduce the cost of event enrichment. In each step, we check only a part of the user query. If any of the subqueries cannot be matched, the whole query is not matched, and the EPA sends the event to the event sink.

The givens for the optimization problem are a user query, the raw event stream (including event types) and heuristics about the external KB. We require for an optimized execution plan a user query with acceptable latency and costs (computation, materialization, and network transmission costs).

We can assume that the events can be kept in the same order as they arrived at the base point of the system; the time distances between event instances and the background knowledge are not significantly changed while an event instance goes through the multiple processing steps. Multistep event enrichment and detection can be planned to optimize the processing costs.

Planning Multistep Event Enrichment and Detection

Multistep event enrichment and detection approach (shown in Fig. 10.3) consists of several steps of event enrichment followed by event detection steps. In each of these steps, a part of the event enrichment is executed, so that event detection can be realized based on the added knowledge of the enrichment step. The two main tasks for multistep processing are to be defined for each processing step: a knowledge query that can extract the required knowledge from the external KB for event detection without missing any of the event components, and an event detection pattern based on the enriched knowledge for detection. For the optimization of the processing steps, we need to set up subgraphs at the initial steps which are less likely to be matched, so that the downstream discharge of the processing steps can be minimized and costs for the event enrichment optimized.

10.5 Summary and Discussion

In this chapter, we have described the main concepts of SCEP. We described the different existing approaches for semantic event processing and presented the approach for semantic enrichment of event streams advantages and disadvantages of event enrichment. Furthermore, we provided a planning approach for the optimization of enrichment costs, so that the system can detect complex events within user-defined latency expectations and, on the other side, optimized event enrichment costs. In order to generate different enrichment plans, the given user query is preprocessed, and several subgraphs enrichment of events are generated.

The application of SEES in event processing use cases can be realized considering the advantages and also disadvantages of this approach.

Advantages of SEES:

Distribution of Processing Task (Higher Scalability) The main advantage of SEES is the capability of distributing the task of event enrichment to multiple EMA, which allows better scalability and overall performance of the SCEP system.

Possibility for Knowledge Modularization One further benefit of task distribution is it allows for modularization of domain background knowledge to multiple knowledge modules. Each EMA can use some of the KB modules.

Including Update Knowledge As described, the external KB might have some knowledge updates (ABox updates) with clearly lower update frequency than the main event stream. One advantage of semantic enrichment, compared to other approaches, is extracted knowledge from the KB, which is the latest version of the available knowledge, so that the latest updates can be used for the enrichment of events. In the case that the knowledge changes the event processing engine can include these updates in the event detection process.

Disadvantages and Problems of SEES:

Large Number of Derived Events The main disadvantage of the semantic enrichment of events can be the management of huge amounts of generated new events which are produced by EMA. These events should be processed by the final CEP engine to match the complex query. From a single event, instances of several new events are generated, or several new attribute fields are added to the event attributes (fat event enrichment). Only a part of these derived events are used at the end to match the complex events and the rest are moved to the event sink (because they are only generated for internal usage, that is, unnecessarily derived events are deleted without any usage). The SEES approach can scale while lots of derived events are unnecessarily produced. According to planning approach, an optimal plan for event enrichment and detection reduces the costs of event processing.

Enrichment Latency The enrichment of the event stream prior to the event detection process increases the latency of the CEP system, because in the first step the system has a latency for the enrichment of events, the results of which are transmitted to the second step for event detection.

Considering all the assets and drawbacks of the SEES approach, we can conclude that this approach is applicable for use cases of knowledge-based event processing where the scalability of event processing is an important factor; latency expectations allow a prior processing step for the enrichment of events, thereby reducing enrichment costs.

References

Abadi, D. J., Carney, D., Cetintemel, U., Cherniack, M., Convey, C., Lee, S., & Zdonik, S. (2003). Aurora: A new model and architecture for data stream. *The VLDB Journal*, *12*(2), 120–139. doi:10.1007/s00778–003–0095-z.

Agrawal, J., Diao, Y., Gyllstrom, D., & Immerman, N. (2008). Efficient pattern matching over event streams efficient 2pattern matching over event streams. Proceedings of the 2008 ACM SIGMOD International Conference on Management of Data (pp. 147–160). New York: ACM. http://doi.acm.org/10.1145/1376616.1376634 10.1145/1376616.1376634. Accessed Jan 2012.

Anicic, D., Fodor, P., Rudolph, S., & Stojanovic, N. (2011a). EP-SPARQL: A unified language for event processing and stream reasoning. Proceedings of the 20th International Conference on World Wide Web (pp. 635–644). New York: ACM. http://doi.acm.org/10.1145/1963405.1963495. Accessed Jan 2012.

Anicic, D., Fodor, P., Rudolph, S., Stühmer, R., Stojanovic, N., & Studer, R. (2011b). ETALIS: Rule-based reasoning in event processing. In S. Helmer, A. Poulovassilis, & F. Xhafa (Eds.), *Reasoning in event-based distributed systems* (vol. 347, pp. 99–124). Berlin: Springer.

Barbieri, D. F., Braga, D., Ceri, S., & Grossniklaus, M. (2010). An execution environment for C-SPARQL queries. Proceedings of the 13th International Conference on Extending Database Technology (pp. 441–452). New York: ACM. doi:/10.1145/1739041.1739095.

Bolles, A., Grawunder, M., & Jacobi, J. (2008). Streaming SPARQL extending SPARQL to process data streams. Proceedings of the 5th European Semantic Web Conference on the Semantic Web: Research and Applications, ESWC'08, (pp. 448–462). Berlin: Springer.

Brenna, L., Demers, A., Gehrke, J., Hong, M., Ossher, J., Panda, B., & White, W. (2007). Cayuga: A high-performance event processing. Proceedings of the 2007 ACM SIGMOD International Conference on Management of Data (pp. 1100–1102). New York: ACM. doi:10.1145/1247480.1247620.

Bry, F., & Eckert, M. (2007). The language XChangeEQ and its semantics rule-based composite event queries. Proceedings of First International Conference on Web Reasoning and Rule Systems. Innsbruck, Austria (7–8 June 2007).

Calbimonte, J. P., Corcho, O., & Gray, A. J. G. (2010). Enabling ontology-based access to streaming data sources. Proceedings of the 9th International Semantic Web Conference—Volume Part I. Berlin: Springer. http://dl.acm.org/citation.cfm?id=1940281.1940289. Accessed Jan 2012.

Chakravarthy, S., & Jiang, Q. (2009). *Stream data processing: A quality of service perspective modeling, scheduling, load shedding, and complex event processing* (1st ed.). Berlin: Springer Publishing Company, Incorporated.

Chakravarthy, S., & Mishra, D. (1994). Snoop: An expressive event specification language for active databases. *Data and Knowledge Engineering, 14,* 1–26. doi:0.1016/0169–023X(94).

Chakravarthy, S., Krishnaprasad, V., Anwar, E., & Kim, S. K. (1994). Composite events for active databases: Semantics, contexts and detection. Proceedings of the 20th International Conference on Very Large Data Bases (pp. 606–617). San Francisco: Morgan Kaufmann Publishers Inc. http://dl.acm.org/citation.cfm?id=645920.672994.

Chandrasekaran, S., Cooper, O., Deshpande, A., Franklin, M. J., Hellerstein, J. M., Hong, W., & Shah, M. A. (2003). TelegraphCQ: Continuous dataflow processing for an uncertain world. Proceedings of the 2003 ACM SIGMOD International Conference on Management of Data. New York: ACM. doi:10.1145/872757.872857.

Cranor, C., Johnson, T., Spataschek, O., & Shkapenyuk, V. (2003). Gigascope: A stream database for network applications. Proceedings of the 2003 ACM SIGMOD International Conference on Management of Data (pp. 647–651). New York: ACM. doi:10.1145/872757.872838.

Dean, J., & Ghemawat, S. (2008). Mapreduce: Simplified data processing on large clusters. *Communications of the ACM, 51*(1), 107–113. http://doi.acm.org/10.1145/1327452.1327492 10.1145/1327452.1327492.

Demers, A., Gehrke, J., Hong, M., Riedewald, M., & White, W. (2006). Towards expressive publish/subscribe systems. Proceedings of the 10th International Conference on Advances in Database Technology (pp. 627–644). Berlin: Springer. doi:10.1007/11687238_3_8.

Diao, Y., Immerman, N., & Gyllstrom, D. (2007). SASE+: An agile language for kleene closure over event streams. UMass Technical Report 07-03

Dıaz, O., Paton, N. W., & Gray, P. M. D. (1991). Rule management in object oriented databases: A uniform approach. Proceedings of the 17th International Conference on Very Large Data Bases (pp. 317–326). San Francisco: Morgan Kaufmann Publishers Inc.

Etzion, O., & Niblett, P. (2010). *Event processing in action* (1st ed.). Greenwich: Manning Publications Co.

Forgy, C. L. (1979). *On the efficient implementation of production systems*. Pittsburgh: Department of Computer Science, Carnegie-Mellon University. AAI7919143

Gatziu, S., & Dittrich, K. R. (1992). SAMOS: An active object-oriented database system. *IEEE Quartely Bulletin on Data Engineering*, 23–39.

Gatziu, S., & Dittrich, K. R. (1994). Detecting composite events in active database systems using petri nets. In J. Widom, S. Chakravarthy (Eds.), *RIDE-ADS* (pp. 2–9). IEEE Computer Society. doi:10.1109/RIDE.1994.282859.

Gehani, N. H., Jagadish, H. V., & Shmueli, O. (1992a). Composite event specification in active databases: Model & implementation. VLDB '92: Proceedings of the 18th International Conference on Very Large Data Bases (pp. 327–338). San Francisco: Morgan Kaufmann Publishers Inc.

Gehani, N. H., Jagadish, H. V., & Shmueli, O. (1992b). Event specification in an active object-oriented database. *SIGMOD Record, 21*(2), 81–90. doi:10.1145/141484.130300.

Gehani, N. H., Jagadish, H. V., & Shmueli, O. (1993). Compose: A system for composite specification and detection. In *Advanced database systems* (pp. 3–15). London: Springer. http://dl.acm.org/citation.cfm?id=647416.725348.

Glombiewski, N., Hossbach, B., Morgen, A., Ritter, F., & Seeger, B. (2013). *Event processing on your own database*. In G. Saake, A. Henrich, W. Lehner, T. Neumann, & V. Köppen (Eds.), BTW Workshops (pp. 33–42). LNI, GI.

Group, T. S. (2003). *STREAM: The Stanford stream data manager*. Technical Report 2003–21. Stanford InfoLab.

Gyllstrom, D., Wu, E., Chae, H. -J., Diao, Y., Stahlberg, P., & Anderson, G. (2006). *SASE: Complex event processing over streams*. http://arxiv.org/abs/cs/0612128. Accessed Jan 2012.

Jensen, K. (1996). *Coloured petri nets: Basic concepts, analysis methods and practical use* (vol. 1, 2nd ed.). London: Springer.

Kemper, A., Neumann, T., Funke, F., Leis, V., & Mühe, H. (2012). Hyper: Adapting columnar main-memory data management for transactional and query processing. *IEEE Data Engineering Bulletin, 35*(1), 46–51. http://sites.computer.org/debull/A12mar/hyper.pdf.

Kozlenkov, A., Penaloza, R., Nigam, V., Royer, L., Dawelbait, G., & Schroeder, M. (2006). Prova: Rule-based Java scripting for distributed web applications. Proceedings of the 2006 International Conference on Current Trends in Database Technology (pp. 899–908). Berlin: Springer. doi:10.1007/11896548.

Kraemer, J., & Seeger, B. (2004). PIPES: A public infrastructure for processing and exploring streams. Proceedings of the 2004 ACM SIGMOD International Conference on Management of Data (pp. 925–926). New York: ACM. doi:10.1145/1007568.1007699.

Le-Phuoc, D., Dao-Tran, M., Parreira, J. X., & Hauswirth, M. (2011). A native and adaptive approach for unified processing of linked streams and linked data. Proceedings of the 10th International Conference on the Semantic Web: Volume Part I, ISWC'11 (pp. 370–388). Berlin: Springer.

Luckham, D. C. (2001). *The power of events: An introduction to complex event processing in distributed enterprise systems*. Boston: Addison-Wesley Longman Publishing Co., Inc.

Luckham, D. C., & Schulte, R. (2011). Event processing glossary—version 2.0. Event Processing Technical Society.

Manegold, S., Kersten, M. L., & Boncz, P. (2009). Database architecture evolution: mammals flourished long before dinosaurs became extinct. *Proceedings of the VLDB Endowment, 2*(2), 1648–1653.

Miranker, D. P. (1987). TREAT: A better match algorithm for AI production system matching. In K. D. Forbus& H. E. Shrobe (Eds.), AAAI (pp. 42–47). Morgan Kaufmann.

Paschke, A. (2007). A homogenous reaction rule language for complex event processing. Proceedings of the 2nd International Workshop on Event Drive Architecture and Event Processing Systems (EDA-PS), CoRRabs/1008.0823. http://arxiv.org/abs/1008.0823. Accessed Jan 2012.

Paschke, A. (2009). Rules and logic programming for the web. In A. Polleres et al. (Eds.), *Reasoning Web 2011* (LNCS 6848, pp. 326–381). http://dx.doi.org/10.1007/978-3-642-23032-5_6. Accessed Jan 2012.

Paschke, A. (2014). Reaction RuleML 1.0 for rules, events and actions in semantic complex event processing. Proceedings of the 8th International Web Rule Symposium (RuleML 2014) (LNCS, 18–20 Aug 2014). Prague: Springer.

Paschke, A., & Boley, H. (2009). Rules capturing event and reactivity. In A. Giurca, D. Gasevic, & K. Taveter (Eds.),*Handbook of research on emerging rule-based languages and technologies: Open solutions and approaches*. IGI Publishing, ISBN: 1-60566-402-2.

Paschke, A., & Kozlenkov, A., (2009). *Rule-based event processing and reaction rules*. Proceedings of RuleML 2009 (pp. 53–66).

Paschke, A., Vincent, P., & Springer, F. (2011). Standards for complex event processing and reaction rules. Proceedings 5th International Web Rule Symposium (RuleML 2011) (pp. 128–139).

Paschke, A., Vincent, P., Alves, A., & Moxey, C. (2012a). *Advanced design patterns in event processing*. Proceedings of the 6th ACM Conference on Distributed Event Based Systems (DEBS 2012) (pp. 324–334).

Paschke, A., Boley, H., Zhao, Z., Teymourian, K., & Athan, T. (2012b). Reaction RuleML 1.0: Standardized semantic reaction rules. Proceedings of the 6th International Web Rules Symposium (RuleML 2012) (27–31 Aug 2012). France: Montpellier.

Paton, N. W., & Diaz, O. (1999). Active database systems. *ACM Computing Surveys, 31*(1), 63–103. doi:10.1145/311531.311623.

Petrovic, M., Burcea, I., & Jacobsen, H. A. (2003). S-ToPSS: Semantic Toronto publish/subscribe system. Proceedings of VLDB 2003. VLDB Endowment.

Schapranow, M., & Plattner, H. (2013). In-memory technology enables history-based access control for RFID-aided supply chains. In A. Giurca, D. Gasevic, & K. Taveter (Eds.), *The secure information society: Ethical, legal and political challenges* (pp. 187–213). New York: Springer. ISBN: 13 978-1-4471-4762-6.

Schmidt, K. U., Anicic, D., & Stüuhmer, R. (2008). Event-driven reactivity: A survey and requirements analysis event-driven re-activity. SBPM2008: 3rd International Workshop on Semantic Business Process Management in Conjunction with the 5th European semantic Web Conference (eswc'08). CEUR Workshop Proceedings (CEUR-WS.org). ISSN: 1613–0073. http://ceur-ws.org/Vol-472/.

Sequeda, J., & Corcho, O. (2009). A position paper linked stream data. In SSN09 [SSN logo] International Workshop on Semantic Sensor Networks 2009 (CEUR-WS.org). http://ceur-ws.org/Vol-522/p13.pdf.

Stonebraker, M., Madden, S., Abadi, D. J., Harizopoulos, S., Hachem, N., & Helland, P. (2007). The end of an architectural era: (it's time for a complete rewrite). Proceedings of the 33rd International Conference on Very Large Data Bases (pp. 1150–1160). VLDB Endowment. http://dl.acm.org/citation.cfm?id=1325851.1325981. Accessed Jan 2012.

Teymourian, K., & Paschke, A. (2009a). Towards semantic event processing. DEBS '09: Proceedings of the Third ACM International Conference on Distributed Event-Based Systems. New York: ACM. doi:10.1145/1619258.1619296.

Teymourian, K., & Paschke, A. (2009b). Semantic rule-based complex event processing. RuleML 2009: Proceedings of the International RuleML Symposium on Rule Interchange and Applications.

Teymourian, K., & Paschke, A. (2014). Plan-based semantic enrichment of event streams. Eleventh Extended Semantic Web Conference (ESWC 2014), Crete, Greece.

Teymourian, K., Rohde, M., & Paschke, A. (2012). Fusion of background knowledge and streams of events. In F. Bry, A. Paschke, P. Th. Eugster, C. Fetzer, & A. Behrend (Eds.) DEBS 2012 (pp. 302–313). ACM

Valle, E. D., Ceri, S., Van Harmelen, F., & Fensel, D. (2009). It's a streaming world! Reasoning upon rapidly changing information. *IEEE Intelligent Systems, 24*(6), 83–89. doi:10.1109/MIS.2009.125.

Walzer, K., Schill, A., & Löser, A. (2007). Temporal constraints for rule-based event processing. Proceedings of the ACM First Ph. D. Workshop in CIKM (pp. 93–100). New York: ACM. doi:10.1145/1316874.1316890.

Wienhofen, L. W. M., & Toussaint, P. J. (2010). *Enriching events to support hospital care*. Proceedings of the 7th Middleware Doctoral Symposium, MDS '10 (pp. 26–30). New York: ACM. doi:10.1145/1891748.1891753.

Willhalm, T., Popovici, N., Boshmaf, Y., Plattner, H., Zeier, A., & Schaffner, J. (2009). SIMD-scan: Ultra-fast in-memory table scan using on-chip vector processing units. Proceedings of VLDB 2009.

Zhang, Y., Duc, P. M., Groffen, F., Liarou, E., Boncz, P., Kersten, M., & Corcho, O. (2012). Benchmarking RDF storage engines. Deliverable D 1.2. Planet Data FP7.

Zhou, Q., Simmhan Y., & Prasanna, V. (2012). SCEPter: Semantic complex event processing over end-to-end data flows. Technical Report, Department of Computer Science, Ming Hsieh Department of Electrical Engineering, University of Southern California.

Chapter 11
Semantic Web and Business: Reaching a Tipping Point?

Eric Hillerbrand

11.1 Placing the Web in Context

As the Semantic Web is a technology, discussions on Semantic Web and business have focused on the technology challenges associated with information systems and the ways semantic technologies can resolve those challenges. By focusing on the technological challenges and ways that semantic technologies can help to resolve those challenges, more significant business opportunities have been missed. Across the blogosphere, there are wild adherents to the possibilities of the Semantic Web, but there are a larger number of skeptics. As one blogger opined, "I've been a semantic web skeptic for years. SemWeb is a narrowly purposed replica of a subset of the World Wide Web…the SemWeb offers a vanishingly small benefit to the vast majority of businesses. The vision persists but is unachievable; the business reality of SemWeb is going pretty much nowhere." (Grimes 2014) The skeptic's criticism points to the absence of a clear growth path forward for businesses in adopting these technologies. For adherents, their rebuttal falls into several, frequently rehashed, buckets.

Rebuttal 1: *It is a communications problem.* The adherents of a disruptive vision for the Semantic Web have not done a good job in explaining the Semantic Web to business so businesses simply do not understand. Technologists in business, the typical recipient of the Semantic Web pitch, struggle to understand the total cost of ownership or projected return on investment for adapting these approaches.

Rebuttal 2: *It is a recognition problem.* The Semantic Web is alive but no one knows it. Here, adherents cite incremental moves that indicate that Semantic Web standards are being adopted in variation by the likes of Google or Facebook. And, as the argument goes, once others recognize what has been happening under the radar, the Semantic Web will explode. Critics downplay these examples and point to these examples as isolated examples of a small group of adherents attempting to

E. Hillerbrand (✉)
Knowledgegrids, Inc., 634 Prairie Ave, Wilmette, IL 60091, USA
e-mail: ehillerbrand@knowledgegrids.com

M. Workman (ed.), *Semantic Web*, DOI 10.1007/978-3-319-16658-2_11

spin a groundswell when those in the broader technical community are oblivious to these changes. In none of the well-known cases are these incremental shifts truly industrial-strength deployments.

Rebuttal 3: *It is a naming problem.* Here, adherents lay claim to a broader set of shifts in thinking and cite victory for the Semantic Web. References to "linked data" or "graph-based data structures" such as Facebook's social graph are claims of Semantic Web growth. The question becomes whether those major uptakes of Semantic Web-like technologies are connected in any way to the Semantic Web or whether these are evolutions occurring from simple business requirements.

These wild debates occur primarily within the technical community. The result is an echo-chamber debate that bears little connection to the nontechnical problems faced by businesses, especially consumer-facing businesses. A solid understanding of the business challenges, not technical challenges, reveals exciting opportunities for semantic technologies that will be far more constructive and beneficial to business. The real pains for businesses today are the challenges of a dynamic business environment, a proliferation of channels to interact with consumers, and desire to be more relevant in more and more contexts, communicating in more and more personal ways.

Business marketers have traditionally focused on "segmented" thinking. A segment is a group of consumers that share one or more characteristics of interest to the marketer. Segment-based thinking grew out of the historical limits of marketing channels. Television, as a primary marketing channel, was a broadcast medium. As a result, marketers were forced to think about marketing communication to groups of viewers. With the rise of digital technologies, there is now the ability to no longer think about segments as large groups of consumers. Instead, marketers can think about segments of single consumers and even fashion means to communicate to each of these "segments of one" in highly personal ways. Mobile technologies have pushed this "segments of one" thinking to consideration of individual consumers in context. Context in this instance is considered real-time consumer behavior defining preference, motivation, intention, and reaction. Location data, captured by mobile phones, provides even a richer source of contextual data.

Marketing to consumers as individuals in a specific context, selecting the right message and right channel, is the marketing challenge facing today's consumer-facing businesses. For today's always-online consumers, there is an expectation that marketing is always available and always relevant. To be always relevant, marketers need to not only send relevant messages but also receive feedback from consumers. In other words, today's marketing is a conversation; an exchange between business and consumer in which each is trying to understand the other, building a stronger and stronger relationship of value and trust. Here lies the opportunity for semantic technologies: enabling a highly complex conversation that relies on more and more difficult-to-understand vocabularies expressed across diverse channels of communication. All with trillions of dollars at stake.

A Review of Traditional Semantic Technology Application to Business

A review of the literature on the Semantic Web and business reveals a surprising disconnect between the capabilities of the Semantic Web and business. In perhaps the first offering on the role of the Semantic Web and business, David McComb, in *Semantics in Business Systems* (2004), highlights the inherent difficulty with business semantics, and its impact on business systems. For McComb, semantic definition of relevant data or business processes using ontologies resolves many of the costly challenges in integration of business systems. Business systems reflect considerable semantic complexity and that complexity is evident whenever business systems are integrated. Thus, semantic approaches ease business systems integration by enabling a more flexible and scalable binding of information systems.

David Siegel, in *Pull: The Power of the Semantic Web to Transform Your Business* (2009), adopts a different vision for the role of semantics in business. For Siegel, the power of semantic approaches rests with a shift from a paradigm of search to a paradigm of information "pull." Siegel sees this shifting information access as enabling new types of engagement between consumers and business. "When we pull information, we automatically get what we need when we need it" (p. 11). The pull world is a world of connected data with linkages defined through a semantic markup. Siegel's specific focus is on the notion of a future for consumers of an online digital data locker that contains bits of the Semantic Web: tags, words, and numbers in standardized semantic formats. The forcing function for the rise of the Semantic Web and the adoption of business of Semantic Web technologies is tied to new consumer applications that integrate, from across the Internet, the multiple, fragmented, and redundant data silos of personal information. Siegel contends that "power tagging" a process of marking up real-time information—facts or events from the real world—is where opportunities lie for Semantic Web technologies.

The academic world has increased its own sophistication in applying the Semantic Web to business. Analyses have been conducted on the applicability of the Semantic Web to business process management and business intelligence (BI). Hepp et al. (2008) argued the business case for a unified view of business processes. The absence of a formalized representation of business processes is a major obstacle in the mechanization of business process management. They further argued that differing semantics are the key challenge to creating a unified view, and the use of semantic technologies provides representation techniques that can address these challenges; thus, they proposed a markup language and architecture to support business process definition and integration.

Berlanga et al. (2012) analyzed the convergence of BI and the Semantic Web. BI derives important business-critical knowledge from data. The absence of a shared semantics across internal and external data sets inhibits the ability of BI applications to easily integrate data for analysis. Semantic technologies enable a shared semantics to be used to integrate data. They concluded that semantic technologies have been infrequently used in the data warehouse community partly because analysis is conducted under well-controlled and structured scenarios. The eruption of XML and other richer semi-structured and unstructured data formats is forcing, according to Berlanga, the data warehousing and BI communities to confront much more heterogeneous and open scenarios than that of traditional BI applications. The result is a strong interest to bringing semantics to the analytical process. As BI operations require integration of more and more disparate information sources, semantic technologies will take on a more important role according to Berlanga.

Others such as Lytras and Garcia (2008) have analyzed the factors that have limited the massive commercialization of Semantic Web technologies for business. Lytras asserted that Semantic Web R&D projects have resulted in a myopic view of the Semantic Web as a panacea for all the "data" and "knowledge" management-related inefficiencies. Lytras and colleagues proposed a stage model of critical steps for a "Semantic Web Engineering Approach"; according to Lytras, to deliver "business" value. They cite the critical knowledge gap in methodologies and practices for the adoption of Semantic Web technologies in a real business context. For Lytras, the failure to map knowledge-oriented performance to business processes has resulted in a failure to appreciate the role of semantic technologies. The Semantic Web can, according to Lytras: (1) support the overall performance of internal business processes and enterprise application integration, tying these processes to knowledge related activities, and (2) enhance organizational networking and business synergies with other business partners or potential market, and various individuals or business "stakeholders."

Lytras identifies a number of key issues faced by businesses that creates in turn opportunity for the development and commercialization of semantic technologies; enabling businesses to design and implement real-world Semantic Web applications and not prototypes of limited functionalities. These include:

1. *The Data Layer*: Given technology investments in applications that "represent" data in the form of database schemas for relational databases and object-oriented or multidimensional databases, a "killer transformation application" is required to annotate data to Semantic Web "standards." From a business point of view, it is incumbent on semantic technology vendors to communicate the value of Semantic Web to the "core data and knowledge" of the business, and introduce ways to measure the return on investment of Semantic Web technologies.
2. *The Semantic Web and Ontological Engineering Level*: Standards, tools, products, methodologies, and best practices are absent for providing a unified engineering approach to Semantic Web programing and applications development. There is a critical gap on training and teaching on Semantic Web technologies.
3. *The Semantic Web-based Information Systems Layer*: Information management tools built for managing the semantic aspects of enterprise applications are required. These technologies include "Semantic Web-enabled" databases, Semantic Web-oriented interfaces, and agents or intelligent systems.
4. *The Business Logic/Intelligence Layer*: Finally, applications are necessary to manage the semantics of business functions that enable the core business including the business logic, the business services, and all the supporting mechanisms

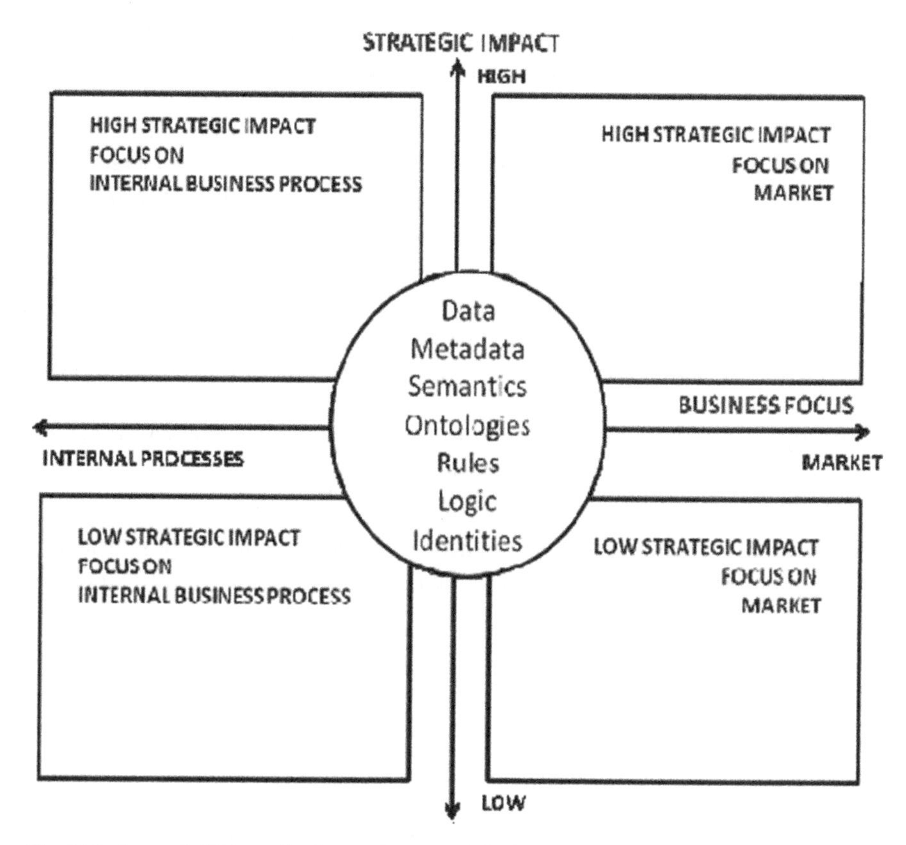

Fig. 11.1 Semantic Web applications: a framework for business and industry exploitation. (Lytras et al. 2008)

that define the "business" in all levels ranging from the daily functional and routing levels to the strategic planning level to the business logic. Lytras frames these layers of application development against an evaluation matrix that propose evaluating semantic technologies based on strategic and business impact (Fig. 11.1).

Lytras' concludes that it is critical to standardize and evolve the tools for the representing of data, managing semantic information such as ontologies, supporting semantic engineering best practices, and extracting and modeling business logic.

Historical Understanding of Business Applications Tied to the Semantic Web

These views of the Semantic Web and business, whether McComb's view that for liberating the pain of business system's integration, Siegel's way of helping businesses reinvent themselves as information brokers, or Lytras' model for developing Semantic Web tools based on their strategic value to the business, reflect a roughly similar understanding of the role that semantic technologies play with the internal side of the enterprise (e.g., functional divisions, partners, and suppliers) and the external side of an enterprise (e.g., customers), which may be either humans (employees or customers) or automated services (e.g., in business processes and enterprise service networks). Semantic enhancement of information delivered to end users and semantic applications can play a role in business to:

- Integrate information from mixed sources through data annotation
- Resolve ambiguities in corporate terminology enabling data integration
- Improve information retrieval, thereby reducing information overload through enhanced search
- Identify relevant information with respect to a given domain through social media analysis and knowledge management
- Provide decision-making support and business intelligence

Several of these functions are reviewed here: (1) annotation of enterprise data, (2) search, (3) social media analysis, (4) knowledge management, and (5) data integration.

Data Annotation

In a business context, annotation enables businesses to find data they need and link data together based on a business need. Semantic technologies aid in data annotation, discovery, and integration. Annotation describes the transformation of syntactically structured data into knowledge structures that represent the meaning of underlying information and allowing the data to be linked together. Annotation defines contextual information that conceptualizes relationships between the data and the real world. Domain conceptualizations, ontologies, or world models provide agreed upon and unambiguous structures for capturing and defining data to which applications, developers, data providers, and consumers can refer. A favorite illustration occurred when working on a Department of Defense development program and asking about whether there was agreement across functional units on the definition of "dead." The answer, after a long pause, was no. "Dead" meant different things to different functional areas in the military, whether supply chain, intelligence, or command operations.

Annotation is an expensive process in terms of human resources. The process also tightly binds the markup to the data in a way that does not support a view that there could be multiple perspectives of a data source especially in different contexts. Annotation of data places a burden on managing multiple ontologies to accommodate the different meanings and needs of different users.

Search

Data discovery, especially as more and more data within the business enterprise is unstructured, places a tremendous burden on search functions. Today, businesses spend considerable time and resources on discovering information. Semantic search engine technology enables searching across multiple data sets including databases, file systems, mail servers, and, in narrow instances, content sources residing outside the enterprise firewalls.

Various data formats such as office documents, pictures, and tables from across every area of the business lead to complexity and, at worst, to information loss. With 80–85 % of enterprise data being unstructured, information cannot be easily browsed with a normal keyword-based search engine. In the case of consumer-facing commerce or content websites, consumers face a similar reliance on search and are equally burdened by its inherent limitations. A semantic search or query system expands the search requests of employees and consumers delivering the exact expected and required results.

Social Media Analysis

The rise of social media, through blogs and social networks, has fueled interest in sentiment analysis. Sentiment analysis has become more significant as businesses try and understand the sentiments expressed by consumers on social media channels, product reviews, or in blog posts. With the proliferation of reviews, ratings, recommendations, and other forms of online expression, online opinion has turned into a kind of virtual currency for businesses looking to market their products, identify new opportunities, and manage their reputations. Looking for a way to understand and measure consumer engagement, sentiment analysis measures the attitude of a consumer towards a brand. For many marketers, the number of Facebook likes and Twitter followers they attract demonstrates their social standing. However, these counts do little to illustrate how consumers are engaging. Through analysis of posts, comments, and suggestions, sentiment analysis provides an understanding of the meaning of what is posted in a format that enables business understanding beyond simple counts. Sentiment analysis uses factors such as context, tone, and emotion and recognition of the subtleties of language, such as sarcasm, or text-speak language such as "OMG" or "LOL." Sentiment analysis provides businesses with an automated way of filtering out the noise, understanding the conversations, identifying the relevant content, and actioning it appropriately.

Knowledge Management

The business world has become so concerned about knowledge management that according to one report, over 40 % of the Fortune 1000 now have a chief knowledge officer, a senior-level executive responsible for creating an infrastructure, and cultural environment for knowledge sharing. Knowledge management is the process of capturing, developing, sharing, and effectively using organizational knowledge. Knowledge management is a multidisciplined approach to achieving organizational objectives by making the best use of knowledge in order to improve performance, competitive advantage, innovation, the sharing of lessons learned, and the continuous improvement of the organization. In order to effectively manage enterprise knowledge, systems are required to classify and categorize knowledge based on a predefined ontological representation. Since much of the knowledge within an enterprise is unstructured or semi-structured, semantic technologies are necessary to effectively organize knowledge for accessibility and reusability.

Data Integration

Data integration using semantics is the process of interrelating information from diverse sources, especially unstructured data such as calendars and to-do lists, e-mail archives, presence information (physical, psychological, and social), documents, contacts (including social graphs), search results, and the advertising and marketing relevance derived from them. Semantic approaches enable the organization of information to correspond to the business objective, and the business requirements for how the information is to be used and accessed. Increasingly, the understanding of data is recognized to be influenced by factors such as context and value.

In the integration of disparate data sources, semantic approaches facilitate and automate communication between computer systems. Even more important, in the face of increasing dynamic business environments, semantic data integration allows greater flexibility in accommodating changes in data, adding new data sources, and changing the mapping between sources as a result of changing business requirements. Semantic technologies use ontologies for creating meta-models of data and business processes. Technologies include tools for ontology management, ontology fusion and mapping, and techniques for ontology refinement through measurement of semantic distance, semantic similarity, and management of mapping rules.

Eventually it is envisioned that semantic-based interfaces to data sources or computer systems can be composed together to enable new and meaningful capabilities. These service compositions would be defined at design or run-time through declarative specifications that are executed at run-time. Although the exact specifications for such systems are largely yet to be defined, the primary methods and means are fairly well understood.

Semantic Web Technologies are Not Keeping Up

While there may be clearer and clearer agreement within the Semantic Web community as to the value that semantic technologies can deliver, that community and evolving semantic technologies confront a nearly universally acknowledged reality:

The World Wide Web was becoming more and more chaotic with the growth in unstructured data. Whether online, through social media, or within enterprise big data conclaves, data chaos prevails. The Semantic Web cannot keep up nor is it likely to ever harness this chaotic universe by bringing order to chaos. Certainly this will not be the case through neatly marked up data based on prescribed World Wide Web Consortium (W3C) standards.

Opportunities may exist in this chaotic future for data brokers to link data sources from across the web into problem-specific webs of interconnected resources that are marked up or stored in some queryable format, selectively findable and accessible via problem-specific tool sets. This is the Mashable.com story: enable integration of data through open application program interfaces (APIs), and for Semantic Web hopefuls, along the way data, is semantically marked up.

The argument that the Semantic Web will tame this burgeoning chaos turns based on conflicting interpretations of the approaches. Is this vision of linked data driven by semantic technologies, or other technologies that bear resemblance to but are not claimed by the Semantic Web community. Is this tamed future a result of semantic markup and auto-extraction of ontological representations of linked data or the use of machine learning and natural language-based systems that beat the Semantic Web to the punch by understanding the meaning of these data linkages. The development of private knowledge repositories, graphically organized with predictive search or semantic query systems are growing: consider, Facebook, Google, or Wolfram Research. Niche providers are developing solutions for businesses that build linked knowledge repositories that tie enterprise, online, and social data together. These developments are more likely the result of the latter techniques than the former. Machine learning, and soft and granular computing techniques, enable the agility required in today's chaotic Internet.

Interestingly, in the face of this trend, Siegel concluded, "My goal was to be the bridge between business decision-makers and SemTech. There's still a huge gap there.... Management seems to be lurching toward [semantic technology in] ways like via social and mobile and Google integration but not via the semantic web].... I got a TON of interest, but no paying clients, so I'm moving on." His focus is now on business agility consulting helping businesses respond to the velocity and variety of the data chaos. "Agile" describes what the Semantic Web is not (Grimes 2014).

Trends that Could Tip the Scale

There are a number of trends occurring within business that are important because of their potential connection to Semantic Web technologies. The opportunities rest with what is happening to businesses and how they relate to their consumers. These opportunities are not driven by Siegel's "pull" transformation but rather the business requirement to drive sales, the more and more complicated understanding of the consumer, and larger and larger amounts of available data. Rather than the "killer app" for semantic technologies, present marketing trends create highly

Table 11.1 Trends that can drive the "tipping point"

Trend	Semantic implication	Technology requirement
Shopper profiles containing data from outside the enterprise	Integration of data with varying semantics	Semantic data integration technologies capable of fusing transaction data with external digital data
Omni-channel, integrated marketing	Highly diverse channel integrations and data requirements	Cross channel ontologies with real-time fusion
Real-time interactions	Highly dynamic, contextual interactions	Contextual ontologies and real-time ontology fusion
Personalized communication with shoppers	Individual representations of "segments of one"	Personal ontology development, extraction, and management

complicated semantic problems that have clear business implications and tangible return on investment.

For marketers, there are four trends that could help tip the scale: First, a focus on building larger and larger shopper profiles containing data from outside the enterprise; second, a focus on omni-channel, integrated marketing; third, a focus on real-time interactions; and fourth, a focus on increasingly personalized communication with shoppers. These four trends all have semantic dimensions as outlined in Table 11.1.

The opportunities presented by these trends focus on building and maintaining ontologies that can integrate heterogeneous data that are dynamic, reformulating those ontologies based on real-time inputs, and executing reasoning based on real-time events and personal shopper characteristics.

Larger and Larger Shopper Profiles

As marketers seek greater understanding of their customers, there is increasing interest in the value of non-transactional data to drive that understanding. Transactional data typically consists of identified shopper purchase histories. For high-involvement retail categories such as grocery, these shopper purchase histories are particularly robust. With an average grocery shopper, shopping trips occur with a frequency of over twice a week. For less frequent purchase categories, data is sparse. Marketers are faced with augmenting these purchases with non-transactional data or data from third parties. Third-party data from data brokers such as Axium, Experian, or Datalogix are highly abstracted data sets consisting, typically, of modeled segment scores. While these are successful businesses, this highly abstracted data delivers little value compared to the power of semantic approaches. This type of data does little to assist marketers in making relevant connections across the diversity of available data. For marketers, there is interest in integrating location data captured from mobile devices, shopper preferences, social network profiles such as "likes" and social interactions that can be easily captured from online behavior.

Additionally, web browsing, search strings, and other captured digital behavior are increasingly available. The result is highly heterogeneous sets of data that reflect not only different semantics at the data source level but also within a source. For example, web page tags that are produced typically by Web developers are not typically managed with a defined ontological point of view. The results are web tags that are wildly divergent in their underlying explanatory power within a single enterprise. These new data sources, regardless of type, are largely unstructured or semi-structured content. So, the marketer faces a tremendously complex semantic integration task with a lack of understanding of the inherent semantic complexities of these data sources.

Omni-Channel Marketing

As shopper profiles are increasing in semantic complexity, the marketer seeks to communicate across more and more marketing channels. Marketer's channels of communication once limited to print advertisements and coupons, and radio and television advertising, have grown dramatically and cross-channel differences are becoming more nuanced. For example, it is now clear that there are often dramatic differences in how consumers are using their smartphones, their mini-tablets, and their tablet computers. Understanding how shoppers engage with a brand across these channels as it leads to a purchase has become incredibly complex. The simple marketing funnel that started with a broadcast message delivered by radio and television to a broad segment of consumers and ending with an in-store purchase is out of date. Today's consumer is bombarded by narrowcast messages delivered digitally and through a mobile device. During the predominance of the traditional marketing funnel, the focus was on a single "moment of truth" or influence, today's environment contains multiple overlapping and nonlinear engagements with multiple moments of truth. A consumer may start with a web search, browse to review sites, return to web search, and respond to a banner advertisement leading to a return to a search engine, all the while interacting with friends through social media.

With a larger number of communication channels and a less controlled marketing process, the costs of maintaining consumer "top of mind" has become a challenge. According to a Bain & Company survey (Customer Retention 2012), an average company loses 20–40% of customers every year. Reducing shopper attrition by 5% can improve bottom-line profits by 25–85%. The stakes are incredibly high. Yet, today's marketing environment may have reached a level of complexity that individual channel management, silos within a broader marketing strategy, will result in failure. Further, the marketing organization may not be able to strategize and manage in such an environment. The marketer may be incapable of reasoning across channels and individual shoppers, responding to real-time events in order to identify "moments of influence." New technologies are critical to aiding in the management of these communications.

Personalization

Personalization is today's marketer's hot button. Forrester Researcher, in a survey of marketing executives in 2013, concluded that marketers believed that personalization was strategically important to their business. Marketers viewed personalization as having the potential to increase traffic, conversion, and average order value. But according to a survey by Monetate (2013), marketers (94 %) know the value of personalization but when data is collected in real-time based on shopper behavior and input, marketers (95 %) reported that they were not able to use the data to deliver in-the-moment personalization.

Most current personalization capabilities use only transaction history. Only 17 % of marketing leaders go beyond basic transaction data to deliver personalized customer experiences. Location-based data, for example, is used by only a small percentage of those surveyed.

Personalization is delivered through a communication channel. Few marketers deliver personalization consistently across channels. Over three quarters of marketers surveyed indicated that they were unable to deliver content or offers to the consumer's chosen channel in real time. Consumers may browse merchandise online and on mobile devices and click to e-mail offers at home, in the office, and on the move. Is there a single in-the-moment personalized communication? Not today.

Loyalty programs are similarly posed for a shift. According to Deloitte consulting, companies that track their customer loyalty are up to 60 % more profitable than those that do not. But consumers are frustrated with the value of the loyalty program to them. The majority of consumers (62 %) join retail loyalty programs so they can get discounts on the things they buy most. Nevertheless, only about one third of American consumers (36 %) receive rewards or promotions that make them come back to the store, and one in four (27 %) of consumers complain they have received reward or promotion for something they would never buy. Eighty-five percent of members report that they have not heard a single word from a loyalty program since the day they signed up. Loyalty programs are leaving consumers feeling underappreciated.

The future of real personalization hinges on technologies that can understand shoppers in the moment, the shopper's location and motivation, and respond accordingly. Only a small percentage of marketers can currently use consumers' real-time location behavior, and intention to drive interactions. Technology constraints inhibit interactions based on what is hot or trending or going on in the moment for that consumer. Semantic technologies that can create a comprehensive, contextually sensitive consumer representations in order to interpret all of the available data in real time are both the challenge and the opportunity.

Real-Time and Contextual Communication

Marketers recognize the value of real-time data in personalization. Marketers are challenged by how to use real-time data to deliver the right content to the right visitor in the right channel at the right time. In the future, marketers will use social sentiment, contextual behaviors, and inferred emotional states in addition to time of day/week and location information. Today, few marketers are able to use inferences about the consumer's emotional state. Similarly, marketers are unable to use individual consumer sentiment to personalize interactions. In addition, few capitalize on the consumer's current context, what the consumer is doing, and the characteristics of the environment that they are doing it in. Environmental characteristics include the presence of competitors, season, day of week, the presence of price and promotion messages, and the presence of friends, experts, or other shoppers that share social affinities. Today, the data that they use to personalize in real time is limited to customer-specified preference data.

Personalization efforts today fail to fully capitalize on—or even ignore—proximity, motivation, intention, and current customer sentiment. Without that understanding, marketers face challenges in delivering real-time, contextually aware, personalized experiences. The technology gap is clear. Marketers lack real-time data analysis and an inability to deliver relevancy to a shopper's preferred channels and in preferred ways.

Success in omni-channel personalized marketing is precipitating a rethinking of customer engagement. Digital technology, and the ability to capture in real-time shopper behavior and input, means that engagement is no longer one way as it was in the broadcast era. Instead, it is possible to engage in the semblance of a conversation. Engagement means message delivery and interpretation of consumer response. A one-way monologue from retailer to customer is no longer sufficient; only a two-way dialogue will enable retailers to understand the shopper and obtain the knowledge of what really matters to them.

Technologies that interpret shopper response and facilitate a true communication in real time between business and consumer is required. In a digital environment in which consumers are always on and expect their social relationships to be similarly available, a business must be capable of not only carrying on a conversation but also must begin to act with the same speed, relevance, and value as an online social network. In the future, the dividing line between shopper's friends and brands that a shopper cares about will become increasingly blurred. The challenge to business is the ability to do this at scale.

The Future of Semantic Technologies and Today's Marketing Challenge

With the focus of today's marketers' consumer-facing business on a new set of marketing challenges, what is the future for semantic technologies? Three opportunities are evident and with those opportunities come new challenges to existing semantic technologies. Marketer's interest in personalizing relationships between brands and consumers in a "segment of one" that is dynamic and contextual will drive these opportunities.

Ontology Extraction

In order to communicate in a personal way with consumers, technologies will be required to reason and predict likely response, and understand consumer response when it occurs since this response will likely be unstructured. Technologies for ontology creation or extraction from a corpus of consumer data will be necessary. As consumers navigate multiple digital profiles of preference, social activity and online behavior marketers will build ontologies that can integrate shopper's profiles. These ontologies will not only integrate actual data but also tie profile attributes to constructs that matter to marketers: behavior on a path to engagement or purchase, motivation, need, interest and preference, and loyalty. Technologies that can extract ontological representations of these "sentiments" and tie these sentiments to profile attributes will be critical.

Personal Ontology Development

These ontological representations will be at the individual level, not at the segment level. That means building an ontology that reflects how a single consumer organizes the world with consumer and the brand in the center of this representation. Consumer segments, this grouping of consumers based on shared attitude or behavior, are presumed to have different definable attributes that affect how and when the consumers in that segment engage with the brand. It is reasonable to assume that these attributes could and should be modeled in an ontology. The presumption is that these ontologies would, in fact, be different across segments. With the focus on "segments of one," the ontologies will move from group-focused representations to individual representations. In other words, systems that manage ontologies for each individual consumer will be required.

Personal Ontology Fusion

As previously discussed, context will become incredibly important to marketers. Context will influence intention and likelihood of engagement. Environmental factors such as location, alternative brands that provide equivalent products or services, alternative brands that provide similar but not equivalent products or services, characteristics of engagement whether communication with the brand or purchase, and channel will interact with individual consumer attributes and in-the-moment factors such as intention, motivation, and behavior. All of these factors, whether environment or in-the-moment shopper behavior and textual input, will be accessible in the form of rich structured and unstructured data that will need to be fused together. The ability to merge together ontological representations and reason successfully about an individual consumer with a specific intention at that moment in a specific location that has certain competitive factors will be required.

Reasoning and Simulation for Consumer

For consumers, the future shifts from a product-focused to need- or experience-focused engagement with brands. Brands that matter will need to be solution focused rather than product focused. The promise of big data to business, and its corresponding hype, has neglected the value to the individual shopper. Here the construct of "small data" has relevance. Small data is the data that individual consumers provide about themselves. Small data in combination with big data is where value can be delivered to individual shoppers. The ability for an individual shopper to benefit from gross and global trends and insights in the context of their individual interests and needs is paramount. Use of massive compute power to help individuals be smarter is a huge promise but it also allows businesses to rethink their value proposition. In the future, businesses may provide not only a product and service to support that product but also data and analysis to help the individual shopper make sense of when to purchase, how to purchase, and then how to use the acquired product.

Sophisticated models of products and individual consumers will enable reasoning and intelligence to benefit not only the consumer but also the business. Simulation or proscriptive analysis will improve the relevance of marketing and have considerable impact on creating an emotional loyalty (not behavioral loyalty as presently exists) between business and consumer.

Agent-Based Frameworks

Retail started in a marketplace in which a known proprietor acted on behalf of the consumer by selecting the right goods. This process of "retailer acting, as agent" continued well into the twentieth century and only changed with the emergence of the grocery store and full-line department store. Retail changed from a focus on agency to a focus on utility and efficiency. Shoppers assumed the role of stock picker and were put into the position of having to serve themselves. Retailers presumed, and the market confirmed, that shoppers appreciated severing the ties to a human resource intensive model. Once endorsed by shoppers, retailers were able to achieve significant savings in staffing.

The focus on increased personalization extends beyond marketing to actual service, and there is increased recognition that the role of retailers (especially physical or offline retailers) needs to change. Retailers will need to return to the model of acting as shopper agents whether in finding the right product, at the right time, at the right price, and the right relevance for the individual shopper. This will need to be done at scale: for millions of shoppers. The shopping process online, already difficult, will become worse as choices proliferate and globalization expands purchase options. While smaller in scale, physical retailers struggling to be relevant and trying to reinvent themselves will make in-store experiences more engaging and personal; all at scale.

Agent-based software frameworks will allow retailers to scale their newfound position as shopper agent. These frameworks will use knowledge of individual shopper historical behavior and preference combined with in-the-moment shopper intention and motivation to navigate product discovery, personal pricing, promotion, purchase, and service. Agent reasoners can project future service and product requirements, and assist customers with optimizing their purchase with upsell/cross-sell or social commerce engagement.

11.2 Conclusions

The Semantic Web lays out a utopian vision. Semantic technologies have continued to evolve. Business applications of semantic technologies, and adoption of the Semantic Web, continues to falter. This is due in part to the background of Semantic Web proponents: technologists or academics. These backgrounds have a limited vision for semantic technologies to primarily technical problems or to highly fantastical visions of futures. In both cases, the return on investment and value to business today is highly suspect.

From the business side, businesses face very real challenges as they continue to respond to the digital age and the age's impact on consumers. All of these challenges have a semantic component. Unlike advocating semantic technologies to resolve data integration challenges, a point of view that forces semantic technologies into a standoff with traditional non-semantic technology approaches and competition with major technology businesses that sell non-semantic-based technologies, these newer challenges to business are ripe for semantic-based solution.

The challenges are related to how to improve consumer understanding, how to understand shopper behavior in the moment, and how to respond in highly personalized ways. The chaos of today's digital environments requires strong approaches for understanding data that is unstructured and events that are contextual, personal, and are complicated to understand.

The opportunities for semantic technologies reside not with the technology side of the organization but with the marketing and merchandising sides of the organization. Here is the lifeblood of the business and the disruption in loyalty, marketing, and product development caused by the Internet, creates a perfect opportunity to deploy new technologies that have immediate return on investment and can speed up the adoption of semantic technologies.

References

Berlanga, R., Romero, O., Simitsis, A., Nebot, V., Pedersen, T., Abello, A., & Aramburu, M. (2012). Semantic Web Technologies for business intelligence. In M. Zorrilla, J.-N. Mazón, Ó. Ferrández, I. Garrigós, F. Daniel, & J. Trujillo (Eds.), *Business intelligence applications and the web: Models, systems, and technologies* (pp. 310–339). Hershey: Business Science Reference.

Customer Retention, Repeat Sales, Referrals– are Business Imperatives: Bridging with Customer Satisfaction, Loyalty, Lifetime, Score (2012, March 18). http://bizshifts-trends.com/2012/03/19/customer-retention-repeat-sales-referrals-are-business-imperatives-bridging-with-customer-satisfaction-loyalty-lifetime-score/. Accessed 15 Dec 2014.

Forrester Research. (2013, November 1). Delivering new levels of personalization in consumer engagement a guide for marketing executives: Strategy, capabilities, and technologies required for delivering effective personalization to consumers across channels. http://www.sap.com/bin/sapcom/en_za/downloadasset.2013-11-nov-21-22.delivering-new-levels-of-personalization-in-consumer-engagement-pdf.html. Accessed 1 Dec 2014.

Grimes, S. (2014). http://www.informationweek.com/software/information-management/semantic-web-business-going-nowhere-slowly/d/d-id/1113323?image_number=1. Accessed 11 Jan 2014.

Hepp, M., Leymann, F., Bussler, C., Domingue, J., Wahler, A., & Fensel, D. (2008). Semantic business process management: Using semantic web services for business process management. http://dip.semanticweb.org/documents/Hepp-et-al-Semantic-Business-Process-Management-Using-Semantic-Web-Services-for-Business-Pro.pdf. Accessed 1 Dec 2014.

Lytras, M., & García, R. (2008). Semantic web applications: A framework for industry and business exploitation—What is it needed for a successful semantic web based application. *International Journal of Knowledge and Learning, 4*(1), 93–108.

McComb, D. (2004). *Semantics in business systems: The Savvy managers guide*. San Francisco: Morgan Kaufman.

Monetate. (2013, June 21). From big data to big personalization. http://content.monetate.com/h/i/12311817-from-big-data-to-big-personalization. Accessed 1 Dec 2014.

O'Brien, D. (2014, September 12). CMO insights on omnichannel personalization in retail—brand quarterly. http://www.brandquarterly.com/cmo-insights-omnichannel-personalization-retail/. Accessed 14 Dec 2014.

Siegel, D. (2009). *Pull: The power of the semantic web to transform your business*. London: Penguin Books.

CPSIA information can be obtained at www.ICGtesting.com
Printed in the USA
BVOW10*1119130915

417749BV00001B/2/P

9 783319 166575